Discerning Prometheus

Discerning Prometheus

The Cry for Wisdom
in Our Technological Society

Robert A. Wauzzinski

Madison • Teaneck
Fairleigh Dickenson Press
London: Associated University Presses

Associated University Presses
440 Forsgate Drive
Cranbury, NJ 08512

Associated University Presses
16 Barter Street
London WC1A 2AH, England

Associated University Presses
P.O. Box 338, Port Credit
Mississauga, Ontario
Canada L5G 4L8

The paper used in this publication meets the requirements of the American National Standard for Permanence of Paper for Printed Library Materials Z39.48-1984.

Library of Congress Cataloging-in-Publication Data

Wauzzinski, Robert A., 1950–
Discerning Prometheus : the cry for wisdom in our technological society / Robert A. Wauzzinski.
    p. cm.
Includes bibliographical references and index.
ISBN 0–8386–3866–X (alk. paper)
    1. Technology—Philosophy. I. Title.

T14 .W33 2001
303.48'3—dc21
                                                                00–049033

PRINTED IN THE UNITED STATES OF AMERICA

To Don:
A Virgil for this Dante

Contents

Acknowledgments

Interdisciplinary scholarship must be a joint endeavor. Therefore, I thank the following people for their assistance in the preparation of this manuscript. Professor Ken Herman must be thanked for his historical and philosophical insights. Dr. Edwin Olson contributed his engineering, scientific, and, above all, his editorial insights to the manuscript.

I am fortunate for having been tutored by some who have passed my way who have been labeled students. Hence, I must thank Jack Harris for adding his budding editorial and research skills to this work. Timothy Welman, prisoner of Jubilee, must be thanked for his editorial suggestions that will help many students in years to come. I especially want to thank Kristi Marden for her typing and editorial skills displayed in the bibliography and Helene A. Hoover for her thorough editing of the entire work. These talented students contributed to some of the strengths and none of the weaknesses of this work.

Thanks also must be offered to members of the library staff of The Free University of the Netherlands for assisting my research. Their command of English greatly exceeds my faltering Dutch.

Above all, thanks must be extended to Professor Schuurman for suggesting resources that filled some gaps in my research.

Foreword

Discerning Prometheus represents an interesting and provocative contribution to the philosophy and ethics of technology, as well as to the philosophy of culture. Professor Wauzzinski follows nearly fifty years of professional philosophical and ethical interest in technology by reflecting on the meaning and the importance of technology. That this reflection should take place now is understandable, even mandatory, because of the imposition and threats posed by modern technology. Technology has enriched our culture. At the same time, it has produced bigger and more costly burdens. Indeed, technology seems to threaten our increasingly interdependent global culture. The use of nuclear weapons and global pollution are examples of that threat.

The philosophical discussion of technology has been dominated by those having optimistic or pessimistic appraisals of the impact and the meaning of technology. Wauzzinski analyzes these positions in depth and with finesse and nuance. Part of Wauzzinski's originality, however, stems from his development of a typology that analyzes realism and structuralists as well. He builds to his discussion of realism by showing how optimists promote technology without being sufficiently aware of technology's long-term inherent dangers. They expect technology to enhance life through the scientific and technological subjugation of nature and concomitant material rewards but are not sufficiently aware of the negative results of technology. Pessimists, on the other hand, understand all too well the evils inherent in technology. Fundamentally, technology robs modern humanity of freedom. This is ironic when we stop to consider that technology is promoted to enhance freedom of choice. Pessimists say the problems presented by modern technology are insuperable.

11

The realist attempts to maximize utilities and reduce the potential or real harms presented by technology. Realists succeed in this task to a limited degree. However, they do not care about the meaning of technology. They take technology as a pragmatic, factual given. This fundamental problem leads to a form of pragmatism that offers little principled reflection and, hence, little macro control of technology. Their pragmatism seems to offer only a moderate intellectual bridge between the extremes of optimists and pessimists because they cannot find a fundamentally different view of technical reality. The realist's fundamental assumptions differ little from those of the optimist or the pessimist. These assumptions, therefore, become the object of Wauzzinski's penetrating study.

The fourth position is that of the structuralist. Structuralists want humanity to take a holistic approach in relating technology to nature and culture. The ideas of responsibility and normality play a large part in their thinking. Technology is inherently good but often distorted, according to the structuralist. Distortions arise when the meaning and the limits of technology are forgotten. The meaning of technology can be located, in part, in its "enabling" other areas of life to experience more possibilities but, at the same time, not dominating those areas it seeks to enhance, says the structuralist. In short, technology may not dominate life but must enhance it according to principles strucuralists will develop.

Wauzzinski makes both a broad and a deep contribution to the discipline of the philosophy of technology. He is broad in that his interdisciplinary tact covers considerable terrain. His use of the philosophy of economics especially helps him to begin to discern the relationship between economics and technology, a subject only marginally tackled in either discipline. He delves to the depths of the discipline through his discussion of the "roots" or the religious issues at hand. He argues correctly that we may not abstract, à la positivism, the religious-historical background from practical technological development.

Professor Wauzzinski writes with nuance, clarity, and care. His well-documented work does not hinder his style. One gains the impression that he can complement his breadth with precision, which should satisfy professionals in the field while the practicality of the work should find a receptive student audience. His attempt to speak to both the student and the professional is daring and, I believe, successful. The glossary will help the student with the boldfaced terms in the text.

His constructive critique of my position reveals a profound problem now facing many industrialized countries. Professor Wauzzinski rightly maintains that my position is somewhat weak because I do not develop concrete alternatives to problem technologies. I have attempted this in my analysis of modern industrial agriculture. However, concrete alternatives to problematic industrial agriculture can only be heard because of crises that now threaten the very structure or fabric of agriculture. This perceived danger and the willingness to entertain alternatives in other areas of life are not real at this moment. Therefore, structural alternatives are not possible. Sadly, we often do not create solutions for technological problems until crises threaten. Again, the threat posed by nuclear waste comes to mind as an example of our passivity.

Having said this, a note must be sounded clearly. Professor Wauzzinski does the reader a favor when he stresses the need for practical, ethical alternatives to modern technology. It takes courage to couple the study of philosophical thinking with practical alternatives, for it is easy to be misinterpreted by many who do not favor this way of thinking. The book is refreshing in that it opens new theoretical as well as ethical vistas for how technology should service our lives. May many readers accept this challenge to consider more responsibly creative alternatives to modern technology.

I hope that students who are willing to think about our "technological culture," its problems, and its desperate need for principles and direction, will take notice of this important book that addresses some fundamentals of our Western society. We have all been sensing that perhaps we are out of control because our technology dominates us. In many cases our technology has destroyed or will soon destroy our human and natural environments. Too many large Western inner cities testify to this reality. How we use and how we develop technology responsibly are the important questions Professor Wauzzinski begins to raise in this book. We can only hope for responsible answers.

I recommend this book wholeheartedly.

Egbert Schuurman
Breukelen, The Netherlands 2001

Introduction

The headline reads

UNABOMBER A HERO TO SOME

> He's a high-tech Robin Hood. He's going to save us from ourselves.
> There are plenty who feel the same way about technology.[1]

The article concludes that many Americans have mixed feelings about
the "Unabomber." They do not like the fact that he allegedly killed
people, but Theodore Kaczynski apparently speaks for many when he
vilifies modern technology for its destruction of the human and the
natural environment. "Technology must be checked" is the message
Kacyznski sounded that spoke to many Americans. This, they believe,
needs to be said. How did we get to such a level of frustration? Are
there other equally strong, but opposite, feelings about technology? Is
there a middle position between pessimistic and optimistic views
about technology? These are some of the questions that will be addressed in this book.

Exploding space shuttles, congested traffic, spilled poisonous chemicals that cause evacuations of entire towns, and computers outwork-
ing our brains on key activities are but a few of the major
developments and problems caused by modern technology. These
problems and the technologies that caused them have forever changed
our lives. Are the changes brought about by modern technology beneficial for personal and cultural life? Has technology done more harm
than good? Do technological changes bring about both harm and
good? Or is this entire way of thinking about good and bad, harms
and benefits, as important as it may be, the wrong basic question to

ask? Is not the place of technology within our lives the most important issue? These questions must be kept in mind while reading this book. The thinkers I am about to study say answering these questions will change our lives forever.

This book uses different representative metaphors to show how fundamental assumptions affect how we evaluate modern technology. Optimists prefer to think of technology as the universal liberator—a messiah, if you will—of humanity. This charge is deeply ironic because the claim of optimists centers on their treating technological objects as value-free tools. In this case, the fundamental assumption is the autonomy of humanity. On the other hand, pessimists think of technology as a kind of modern Frankenstein's monster. Perhaps not so ironically, pessimists also share the fundamental assumption of autonomy. Realists will argue, in a seemingly more reasonable manner, that technology is laced with positives and negatives. Our job is to measure scientifically the good and bad of technology, argues the realist. Thus, the optimal metaphor seems to be that of the scientist presiding over precision scales that measure out exact amounts of happiness. The fundamental assumption of the realist is a belief that science can control technology and thereby turn its fruits into benefits. Finally, two thinkers—Egbert Schuurman and Ernst Fritz Schumacher—talk about how and why technology should be adapted to the many needs of humanity and to the natural environment. For them, technology is like one room of a many-roomed home. These thinkers I will call structuralists. Their fundamental assumption is that technological development should be guided by principles or rules that do not originate in the autonomy of humanity.

This book will show that there are at least four different ways of viewing the meaning of technology. The term "meaning" has to do with purpose, intent, and identity. To discuss the meaning of technology is to discuss its purpose, intention, or, more generally, its place in our lives. The discussion of meaning will show that distinct ways of viewing technology emerge. These distinct ways are called *types.* A typology is the study of types or ideal and representative ways of thought.

In this book, I will show how technology affects our personal and social lives because it is rooted in a deeply profound reality. I will argue that basic or foundational commitments affect the place technology occupies in our personal and social life. If it occupies a large place, fueled by human autonomy, then we may say that technology's meaning is both broad and deep. If we want no part of it, because it represents a deterministic evil, we may say that it should not be important and, hence, should have a greatly reduced meaning. So it is with only optimists and pessimists: their respectively rosy and gloomy outlooks on life help determine the place that technology takes within life. Plumbing for an understanding of our commitments is a valuable exercise because it will reveal the connections between foundational principles and practical policy outcomes. This is true for realists and structuralists as well.

I am doing science in both a broad and a specific sense in this book. Broadly speaking, science is focusing on a delimited disciplinary field. The philosophy of technology is the guiding and delimiting discipline of this study. All science has at least two tasks: analyzing and classifying. First I will analyze different positions and their main values to determine similarity and composition of thought. Next, I will classify or group these similarities into four recognizable views.

I am attempting an analysis of foundational or inner principles and relationships. What are the important principles that form the four types? Autonomy or self-law is one such important underlying principle. Are we, as the poet claims, "the master of our fate, the captain of our soul"? Or is the task of technology motivated by the rational reconstruction of reality? Relatedly, is freedom the core virtue that motivates us to escape all external hindrances? Or are we pleasure- or utility-calculating machines, as the realist would suggest? Or are we directed by God to respond in a certain way that blends the necessary demands of technology with those of other areas of life? These are the crucial questions that will occupy this book, which must be kept in mind as I analyze crucial underlying principles. Finally, an analysis of these inner principles and relationships should lead us to consider the relationship of technology to personhood: Are humans to be defined as *Homo faber* or man the tool-maker? All the positions studied will assume, sometimes tacitly, a philosophical view of the person.

I am arguing that any practical technology is formed on the foundation of foundational assumptions. These assumptions I will call "religious," only in the sense that they are preconditions for reflection and practice. I am not saying that any particular sectarian position is being discussed, only that basic commitments are at work. Ignoring or minimizing the connections between any given technology and its foundational principles promotes a superficiality and could lead to technological harm. A case in point is our development of nuclear technology. Under the guise of "making ourselves lord and master over nature," and under the principle of progress, we have optimistically developed nuclear energy. The fact that we, as a nation, have not collectively taken time to reflect on this foundational connection, running instead willy-nilly into the development of "atoms for peace" before our wisdom matched our technical knowledge, has led us to an economic, technical, and social nightmare because we cannot safely dispose of the spent fuel. Deep thought takes patience as does the requisite public debate, but these costs are much easier to bear over time when compared to the costs engendered by ignorance.

This connection between practical technologies, ideas, and religious principles forms the foundation for an interdisciplinary methodology. While I have argued for disciplinary focus, I am arguing against a myopic overspecialization. Accordingly, one's necessary disciplinary focus too often obscures other aspects and disciplines and their data. Therefore, to philosophy I add ethics and its analysis of foundational principles, the history of science and technology because of the evolution of technical practice, economic history, and the philosophy of religion. I take this tack not to be intellectually chic, but because **ontological** and intellectual reality demands it.

Thus, I am attempting a foundational analysis of technological thinking. At the root or foundation of any given view or practice of technology a core conviction can be discerned. This conviction, such as freedom, leads to a relationship with practical technologies. This relationship may be one of repulsion or one of adhesion or both, but it is not neutral. Humanity develops technology and, thus, must come to articulate its view of itself in relationship to technology. Are we essentially, at our roots, "tool makers"? The sui generis of the human condition, as well as the relationship of technology to the rest of life, concerns me deeply in this study.

Views are not just sterile, abstract intellectual concepts. Views represent perspective; principles congeal to form lodestars. Perspective is for walking or, more appropriately, for developing. Perspective gives us advice about the relationship of technology to any given problem or area of life. Properly discerned principles keep us on track. Analysis of perspective is foundational analysis.

No house, to continue the building metaphor, is complete without "rooms" in which to live. Rooms, or other areas of life, themselves are the subjects of ontological investigation. This kind of search, as it will be used here, is a search for place and structure. Place means a definable location within life. Technology must occupy a place if it is not to become totalitarian or take up all space. Structure determines the limits and the identity of that place. Structural analysis of technology leads us to the limits, identity, and meaning—hence the structure—of technology and its relation to other areas of life. To make clear this largely implicit relationship is to add nuance to our foundational and structural analysis.

This foundational analysis will attempt to put to rest what is surely one of the most illusory technical myths invented: Technological practice generally and tools in particular are ideologically neutral, or value-free. The tooth-fairy world of disembodied tools and denuded technological practice and artifacts has suffered the same reality shock experienced by all positivism: its foundations are all but destroyed. Particular tools are made by humans who are shaped by foundational assumptions who, in turn, shape or fabricate nature into planned ends. This matrix fashioned on a wide scope may be called culture.

Reality is far more integral than our disciplines allow us to see. At least this is the view that I will argue in this book. I do not think about the chemical, psychological, economic, political, social, and lingual aspects of each and every activity that engages us; the cohesion of reality is taken for granted. Only in analysis is that cohesion fragmented. While temporarily necessary, fragmentation must give way to cohesion unless we want fragmentation and irrelevancy as a permanent legacy.

As I have said, this study is about technological optimism, pessimism, realism, and the structuralists. Optimists believe that if knowledgeable technology is applied in expanding amounts to perennial human problems, such as hunger, these dilemmas will lessen if not vanish. Conversely, pessimists locate the origin of modern evil and problems in technology. Realists talk about both good and bad in technology with the optimal question being, "what are we willing to trade off to secure a safer, more risk-free use of technology?" Finally, structuralists want to talk about structures being decisive for the discussion. Structures are lawful patterns of order that give life limits, meaning, texture, and goodness. Multifaceted patterns require the development of technology within a holistic context for life. The development of the notion of a holistic context represents the singular contribution of structuralists to the typological viewing of technology. Problems arise when this holistic context is tortured, argues the structuralist. Technology must be adapted to enhance, not dominate, life.

Students have suggested that this book be written. It has been my privilege to teach an interdisciplinary course on technology, at different times over thirteen years. During that time many students have helped shape and revise the concepts contained in this book. That said, as research for the book was begun, there appeared to be a hole in the professional literature. No work that surveys the field with an eye to outlining types of thinking could be located. Egbert Schuurman's classic and sweeping tome deals only with optimism and pessimism. Realism and the groundbreaking work of structuralist E. F. Schumacher were not included because Schumacher's work was just beginning to be made available in Europe when Schuurman published his own work.[2] Ian Barbour's ethical survey is more intellectually sweeping, though less thorough, than Schuurman's. Barbour wants to pursue, rightly, a contextual framework for understanding technology. He wants to place technology within an "ethical" context of life. However, because he does not develop an ontology—an explicit view of life's aspects or layers—he can neither place nor survey types of technological thinking because no developed theory of context is present.[3] I, on the other hand, presuppose an integral ontology[4] that will be developed and explained throughout the study. Finally, there are excellent books that deal with individual types of thought. The problem with such pieces is their limited scope.[5]

Thus, the book takes perhaps a dangerous path. Because I attempt to speak to two audiences, both students and my colleagues in the guild, some may charge me with inadequately addressing either group. This concern is addressed throughout the book; I give student and casual readers many practical examples of technology and how they were influenced by the position under consideration. Further, the text does not simply ramble on but reacts in scholarly language to the experts in the field. I attempt to give my own point of view in a clean, crisp manner, but I ask readers to be patient and realize that ideas and technologies have historical legs. They walk to us, as it were, through history and thereby change and adapt. Walking backwards and through seemingly thick jungles of ideas, footnotes, and details will help us find deeper wisdom and insight. I ask my colleagues to consider that there is a relationship between our world of ideas, principles, and the technologies we see around us. This idea contradicts the established wisdom that views technology as ideologically neutral.

That said, it is my hope that this book will stimulate all of us to begin to consider the place technology occupies within our lives. To that end, each chapter contains how, by implication or by direct quote, each type of thought would place technology within life and why such placement is a logical outcome of its assumption.

The term *place* denotes several things. First and foremost it means location in relationship to other things or activities. We presuppose a multidimensional world. Reality is like a layered artichoke, the leaves of which are wrapped around a core meaning we shall soon (with some trepidation) call "religion." In any specific area of life—such as technology—we occupy a place with respect to technology; or, conversely, technology occupies a place within the rest of life. That reality is varied should not surprise people whose livelihood arises from occupying a place within a multidimensional experience we call the liberal arts curriculum.

My assumptions are simple. Technology will neither go away nor dominate our lives. Perhaps, then, we should ask what principles will guide its use and how should it enhance our lives? My survey will address these concerns and probe these questions.

Persons who will be surveyed are chosen for three basic reasons. First, each has made a representative contribution to the discipline by highlighting key issues that need to command our attention. Second, early thinkers deserve to be recognized as pioneers so that those of us who follow may be guided by their grand successes as well as their shortcomings. Finally, the representatives' thinking lends itself to excellent dialogue.[6]

The study of technology needs no more justification than does the study of economics, sociology, history, theology, or any other recognized discipline. This study is made urgent by four different kinds of events. First, disasters like *Challenger*[7] and Chernobyl underline the dangers posed by technology. When the products of technology come crashing down in our midst, or filter into our lungs, it is time to reflect. Second, nationally recognized guilds like the Society for the History of Technology (SHOT) call for an interdisciplinary analysis of technology.[8] Third, the legacies of such disparate figures as Jacques Ellul and E. F. Schumacher demand that we continue the quest for wisdom about the importance of technology for our lives. Finally, I will show that technology shapes the way we view ourselves, the world around us, the nature of right and wrong, the human task, and gods/God.

Several terms are important for our study. Technology is defined differently by each thinker. In other words, one's assumptions about technology will affect one's understanding of the meaning of technology. Not as important but still to the point is my definition of technology. Technology is the powerful ability to fabricate or shape reality by tools to ends that are more practical than theoretical. Humans can shape nature and dramatically influence culture.

This is power. The power to construct culture is part of our nature and represents great promise as well as great threat. We construct tools so that some practical end may come of the task. This apparently has been so since the dawn of humankind. I believe our ability to shape tools is one of the qualities that differentiates us from animals, however much else we share in common. I will show throughout this study that human reason has set before itself the task of dominating nature by redesigning, shaping, and making it a means to the end of human happiness. In its practical, engineering-oriented form, I call this instrumental rationality. The economic fruits of this effort are believed to be bits of utility or happiness that serve as rewards for our efforts.

Philosophy is used as a comprehensive discipline capable of first charting, then plumbing questions of great breadth and depth. To take this path requires an understanding of the ontological warp and woof of reality. Understanding that reality as comprised of patterned, lawful, enduring structures that give shape and meaning to life is crucial for this study.

This view of the layers of reality will lead us to ask related key questions. Does the optimist do violence or justice to demands that exceed technology? Does the pessimist's view of technology obliterate the goodness of reality? It will be crucial to query realists to see if their method of assessment includes a detailed view of a context from which they want to draw their principles for trade-offs. Finally, structuralists must be questioned about their express wish to be holistic. Do they accomplish their task?

Religion is used throughout this work in a standard but potentially misleading manner. Religion here does not mean a narrow, sectarian faith whose aim is imperial and chauvinistic. Rather, it means a fundamental commitment or concern.[9] I call attention to religion for four reasons. First, all the thinkers we will study use religious language. Second, modern philosophy of science talks about commitment,[10] and "steering fields."[11] Third, we academics increasingly talk about interdisciplinary thought without a core discipline or concern around which things and events may cohere. How do we keep the term *interdisciplinary* from functioning like my mother's Friday-night soup pot? She threw everything not eaten during the week into the brew and mixed wildly until everything more or less blended together. We offer the definition of religious commitments, then, as core concerns related to a differentiated view of life because it provides an interdisciplinary framework that gives coherence to a multidisciplinary analysis. Consequently, the prefix *multi-* may be changed to *inter-*disciplinary. Students have likened religion to "a well-spring" and the differentiated view of life to "a many-roomed home."

Because I am looking at commitments, steering fields, and basic issues, I will be doing foundational analyses. As the term implies, we will be looking at those issues and beliefs that form the basic core of the theory and practice of technology. This should not imply a Marxist's view that an analysis of "superstructures" will follow. Rather, in subsequent works I hope to delve systematically into a structural analysis of what an alternative technology might look like. This layered approach to reality will inform this study to a considerable degree.

I trust the title *Discerning Prometheus* is a provocative one. Prometheus was the fictional Titan in Greek mythology who was believed to have stolen fire from the gods and thereby set two important events in motion for humanity: technology and a form of "salvation." Prometheus was punished by the gods, by being bound, for this act of defiance. It is interesting to note, in this regard, that more than one book relates the development of modern technology to the unbinding of Prometheus.[12] By "discerning" I mean collective, ongoing, scholarly, and popular wisdom that attempts to view technology more responsibly and, thus, identify its strengths and weaknesses. Discerning must include but go beyond the use of our empirical senses. It must include the humanistic and social sciences as well as the natural sciences. Perhaps the lack of a national technology policy that clearly articulates deployment and spending strategies along with equally rigorous ethical standards can be located in our collective lack of discernment.

Conversely, the United States has chosen the technological imperative as its operative norm. This standard states that anything that can be done technically should be done as soon as possible. Waste and even grief are the consequences of our obedience to this arduous taskmaster. The shortest survey of the development of nuclear power in this country represents an example of how and why technical and financial abilities can greatly outdistance humanistic wisdom with the consequence of waste and grief. This kind of irresponsibility has led more than one person to pessimism or has resulted in waning support for technology.

Now is the time to begin such collective discernment. The nineteenth century with its optimism has long since ended. As the twenty-first century dawns, our greatest problem is perhaps disillusionment with both the fruits and the fact of technology; at least this is what some "postmodernists" I study will argue. As we begin a new century, perhaps it will be characterized by discernment rather than by optimism or despair.

Finally, a "user-friendly" glossary has been added to the end of the book to aid the student with words that too often seem perplexing, to say the least. The words are meant to be studied while reading the chapters. Words appear in the glossary in the same order as they appear in the text, and are highlighted in the text by bold print. These are not dictionary-tight definitions because different contexts and authors tend to modify the literal meanings of words to meet their needs.

Professional and personal experience has taught me that most people feel ambivalent about technology. We like the way it lifts burdens from our shoulders, the comforts it provides, and the needs it addresses. At the same time, we fear it because our lives and especially many of our jobs, our environment, and personal space are threatened by technical inundation. Only collective wise reflection can deliver us from this anxiety. To that end, I welcome you to *Discerning Prometheus*.

Discerning Prometheus

1

Technological Optimism

EVOLVING OPTIMISM

Reflecting on the meaning of technology requires more than pondering certain technical disciplines. Why? Because technology deeply touches all of life. Therefore, several key disciplines will be mined for their insights. The four views of technology require me to make use of history, philosophy, religion, ethics, sociology, and economics. These disciplines are essential to cultivate because such ideas as human autonomy, the place and nature of reason, the nature of humanity, and humanity's attempt to dominate nature require interdisciplinary thought to show how thoroughly these notions penetrate our lives. It is important to demonstrate not only how basic ideas influence the study but determine the practice of technology as well. Thus, I devote some time to the analysis of such basic ideas as the nature of humanity and the definition of technology.

This chapter will attempt several goals. First we will look at the history of technology as the optimists would look at it. Second, we will look in depth at the notion of "progress" and how it has changed technological development. Next, we will look at the thought of three key optimists: Julian Simon, Karl Marx, and R. Buckminster Fuller. We conclude the chapter with a discussion of the place and a discernment of optimism.

Technology is a human[1] activity that many optimists believe is neutral or value-free. It is not a thing, a belief system, or a tradition. Technological activity produces things such as hammers, computers, and automobiles, as well as our alleged "technological society." However, the activity that has been defined in the introduction as technology is distinct from these objects. For many optimists technology defines the core of human identity. This conviction informs

27

the meaning of human responsibility, the debate surrounding whether or not technology is "**autonomous**," and the nature of the human–machine interrelationship, as well as the definition of technology. All of these vital issues are linked to the unique activity termed technology.

Human life consists of many different activities integrated in our experiences. This integrality may be called wholeness. Diversity in life roughly parallels the variety of subjects studied in a typical liberal arts curriculum, while the search for truth and wisdom should bind one's education together into one more-or-less cohesive activity. Each discipline possesses discernible boundaries giving these aspects limits and identity. Nevertheless, we do not experience this topical diversity as a composite of unrelated building blocks haphazardly stuck together. No, we experience life as more or less whole people. Wholeness means experiencing unity in the midst of diversity. Psychological joy can light up a person's entire being; a chemical imbalance can signal anxiety or depression that in turn can lead to sexual dysfunction. Economic unemployment can cause family disintegration. Life is whole while being diverse. Modern optimists affirm this wholeness but argue that technology is the key to unlocking the potential inherent in the many aspects of life.

Conversely, life seems fragmented when this basic wholeness is lacking. Disciplines seem unconnected. Coherence is missing. Consequently, education seems meaningless. Indeed, fragmentation and meaninglessness plague contemporary education. Much like the industrial assembly line, education is the experience of unrelated isolated specialties called disciplines. By graduation, students are somehow to have constructed a coherent worldview. Optimists would argue that the key to this lack of integrality is the abandoning of technology, the loss of both object and fabricating. This reliance on technology alerts us to questions surrounding the place of technology within our lives.

Technology has a profound influence on the way we view ourselves, the world, God or the gods, the nature of good and evil, and the remedy for age-old maladies. This deep and pervasive influence is possible because life is whole. Optimism capitalizes on this influence.

People make deep commitments and thus attempt to anchor the diversity and the change inherent in life. Something or someone must anchor life so that in the midst of constant change, especially accelerated by modern technology, we can experience firmness, stability, and constancy. I have called these deep commitments "religious."[2]

Again, religion does not mean merely sectarian theological beliefs or activities taking place in sacred buildings, though these are included. Religion is defined as a person's or a culture's ultimate, or depth, commitment, and belief. Religion is that force or that person that motivates all that one does, including the aspects just mentioned. This commitment may revolve around material reality, human autonomy, external human authority, consciousness, sheer instinctual need, or God. Whoever or whatever it may be, all make religious commitments. Optimists are no different.

Autonomy—literally, "self-law"—represents perhaps the predominant article of faith in Western civilization. "Man is the measure of all things" is the credo. Humanity measures all things by our Reason. Reason is the means and the measure of autonomy. This view of Reason has a history. Reason[3] wanted to free itself from the "nonage" or legalized period of immaturity inflicted upon the West by the Christian church. Secular thinkers, inspired by the Renaissance (1300–1600) and deeply influenced by Enlightenment (1700–1800) writings, turned from divine revelation and church tradition to autonomous Reason for the locus of authority, the ground of ultimate certainty. Philosophers thus motivated by autonomy used Reason to throw off heteronomy or the external authority of revelation and church authority. Optimists presuppose this reality.

Autonomy, in turn, led to the attempt to dominate and subjugate nature. Rationality had to remake nature into culture according to the autonomous dictates of the free personality. This process is called technological fabrication, and it happens according to the autonomous dictates of Reason. The subjugation of nature by Reason is pronounced especially in the German idealism,[4] the prominent philosophy of the Enlightenment. It is focused in the particular philosophy of G. W. F. Hegel. Reason was first thought to transcend or rise above reality, and then penetrate it by understanding its mechanical workings through instrumental rationality. Practical technology was thought to follow as the concrete fruit of theoretical and instrumental science. Rational science and practical technology together were to provide a systematic rational product called culture.[5] This production of culture is the beginning of modern optimism.

Nature was to be studied, counted, measured, and molded to the end or goal of manipulation and control.[6] This religiously inspired, rationally directed, empirically oriented process increased in force after the first part of the eighteenth century. Increases brought success as measured by tangible material products and intangible ideals like freedom. Domination of nature meant freedom. We could be free from the forces that had long controlled us! Economic want, medical disease, ignorance, and, increasingly, natural laws such as gravity were to give way to Reason.

This, however, was the eighteenth century. These views had a history traceable to the Middle Ages, the historical beginning of optimism. Improvements in ship construction, such as the lateen or triangular sail, the advent of the swinging rudder, the use of the compass, and the construction of deeper hulls helped improve navigational abilities. These navigational improvements became technological prerequisites for freedom of movement and later of thought. These improvements propelled explorers outward in search of treasure and trade. Increased material abundance resulted from exploration, as did increasing mastery over the sea. Ship improvements, in turn, led to a reduction of the physical stress of onerous rowing. Easier and safer travel increased travel distances, which in turn led to more knowledge and a consequent reduction in ignorance and superstition. In short, navigational technology was among the technical improvements that enabled humanity to realize dreams of freedom.[7] Technological optimism draws great inspiration and manifests a proud spirit because of these and related triumphs of the human mind. Consequently Europe, circa 1300, went through a **Renaissance**, or "rebirth," where a gaze back through Greco-Roman antiquity initiated a proud, active, technical, secular mind-set. This mind-set forms the keystone of optimism.

Early modern navigational technology contributed to a broader worldview in the same way the telescope broadened our outlook, only this time to the entire universe. As a result of the telescope, Western civilization came to view the relationship of Earth to the universe in a mechanical fashion. The universe seemed like a grand machine—orderly, predictable, and impersonal. Galileo's systematic observation of the heavens contributed to the scientific method and the consequent scientific revolution. That Galileo was tried for heresy and told to recant of his views only inflamed the autonomous spirit of secular humanity. Consequently measuring, testing hypotheses and evidence, systematic observation, empirical Reason, and mathematics—the components of the scientific revolution—were to lead many to conclude that reality seemed to operate by mechanistic or machine-like principles or laws.

This mechanistic philosophy flowers in the philosophy of mathematician René Descartes. This early-sixteenth-century genius likened the world's system to a great "machine made by the hand of God," the regularities or laws of which were the product of iron-clad laws of cause and effect. The body's parts and machine "automata" function in a similar manner, according to Descartes. Newton added to this mechanistic philosophy through his formulation of the universal gravitation laws. These laws need not be anchored in a view of an active personal God. A **deistic** view of God replaced a theistic view of God. Deists believed that God made the universe, and then left it to run on its own autonomous impersonal power. Consequently society and culture were ordered by no theistic **heteronomous** principle. Rather nature and culture, it was thought, acted like a grand autonomous machine left it to function on its own. Thus, the claim to personal autonomy was fortified by a mechanistic worldview.

To what end or goal was such a system constructed? Descartes answers:

> it is possible to arrive at knowledge which is most useful in life, and that instead of speculative philosophy taught in the schools, a practical philosophy can be found by which, knowing the power and the effects of fire, water, air, the stars, the heavens, and all other bodies which surround us, as distinctly as we know the truths of mathematics, we might put them in the same way to all the uses for which they are appropriate, and thereby make ourselves, as it were masters and processors of nature.[8]

Belief in Progress

The goal of this mechanical worldview was the creation of a mental system that is both practical in its fruits and powerful in its ability to control the reality it studied. The power was meant to dominate, master, and possess nature. Nature could be dominated because its laws were known (or thought) to be representations of cause and effect forces. If these forces could be known or grasped firmly by a concept, then this mental apprehending could bend laws and forces to our ends. We could not tolerate the thought that these laws of cause and effect determined our destiny. Our desire for autonomy demanded that we manipulate natural laws to our purposes. The manipulation of natural laws was one important rationale for the birth of modern science. Science, in turn, provided the methodology for the technological imperative. This dictate said that everything that could be done *should* be done as soon as technically feasible. Nuclear power is but one example of this technical impulse to be quickly "up and operational." Obeying this dictate was thought to yield total social progress out of which would blossom total human or personal development; so the belief goes for the optimist.[9]

The technological imperative signals the start of a new secular ethical motivation for doing technology. The promise of social and personal betterment—progress—ignited a passion in the human breast. The mandate to manipulate nature technically proved beneficial, as the **Industrial Revolution** would show. This revolution was the culmination of the mechanical philosophy and provided the optimist with a literal world of evidence suggesting that trust in human autonomy and rationality produced tangible rewards. We bent nature to our purposes through the grasp secured by our rational concepts producing, thereby, modern abundance. May technology be praised.[10] Therefore in the successful subjugation of nature the primary rationale for optimism's confidence in technology to produce a better life is located.

The optimist's certainty is based not only on the manipulation of material reality, but on noteworthy achievements in health care and commodity diversification, as well as in safety and security. Manipulation of material reality and consequent product abundance helped define especially the American character. Thus, without apology or embarrassment, American optimists in speaking of the material abundance produced by American technology say, "the material approach of technology fitted admirably the tone of life of a bustling American people who wanted to get things done quickly, who WORSHIPED abundance, and who believed that every free citizen was entitled to a generous share of the things which brought physical comfort in THEIR world."[11]

Technological projects in nineteenth-century America were hailed with optimistic fanfare. As our nation searched for unity during the eighteenth and early nineteenth century, many turned to a belief in the unifying effects of technological objects. Celebrated technical objects and projects represented our ethos and contributed to our civil or our common religion. In 1839 the Reverend James T. Austin echoed the sentiments of a nation when he spoke of the influence of steam power: "It is to bring mankind into a common brotherhood; annihilate space and time in the intercourse of human life; increase social relations; draw closer ties between philanthropy and benevolence; multiply common benefits . . . and religion an empire which

they have but nominally possessed in the conduct of mankind. . . ."[12]
Here is the optimistic spirit in full swagger.

The nineteenth century was an age marked by a belief in the notion of progress. Technical and economic progress became equated with more or less total human and cultural betterment. This equation of technical improvements with human betterment was believed as an article of faith.[13] The problems of want, ignorance, superstition, emotional dysfunction, social dislocation, and alienation were believed to regress as we progressed. With this hope of betterment propelling us, progress became a social imperative that enhanced the technological imperative just mentioned.

The standard of technological progress, then, functions as a secular socioethical principle that gives validation to each technical innovation. The promise of social and personal betterment demanded that we pursue progress with all haste. The evidence for progress seemed to increase with each technological improvement. Secular ethics became tied, therefore, to technical objects. Were not more machines that reduced the strain of human labor being invented? Did not the railroad bridge the vast expanses of the American continent and, thereby, help us fulfill our "Manifest Destiny"? Did not the Industrial Revolution—*the* manifestation of progress, as the quote on the steam engine indicates—signal the beginning of the end of human want? Did not mass production mean more consumption for the masses? Euphoric affirmations to such rhetorical questions elicit faith, not doubt, in technology. The eighteenth century witnessed the blossoming of the long-dormant notion of progress into the cardinal secular imperative. The power of the ideal of progress guided us in our journey toward human improvement. The addition of one hundred years of perceived betterment in the nineteenth century only increased the certainty of the optimist.

I am arguing that progress functions as a secular ethical ideal. It is secular in that a conscious effort is made to derive this standard from sources outside the Bible and and the church. It is ethical in that conduct is measured according to that standard. It is a standard in that it has moral force to sanction or disapprove of behavior. It is a faith because there is an unproven dogma that believes the human condition is improved with each new gadget. Just how much of an uncritically accepted dogma this is will become clear in the chapter on pessimism. Finally, there is a view of history that believes that the human condition is becoming brighter and better in all ways.

For example, the Industrial Revolution—an improvement over eighteenth-century production methods—was fueled by greater power sources that, along with mass production, produced a quantum leap in the volume of goods. Continuing that optimistic trajectory, enthusiasts thought that the historic problem of the lack of abundant energy supplies could be solved or at least significantly addressed by the next century. This was the sentiment behind the advent of nuclear power. Similar historical confidence is demonstrated by modern optimists who argue that computers will one day exceed and replace the functions of human brains. Just as machines enhance and out-perform our muscles, the powerful computer can expand and eventually surpass our thinking processes. Consequently, grand problems such as ignorance, and lesser problems such as the current malaise hanging over public education, will abate. Thus, the good life is believed to originate in technical innovations. The optimist has joined deep autonomous designs with ethical imperatives to produce technical betterment: a potent combination indeed.

The History of the Belief in Progress

The notion of progress is understood better by looking at a historical contrast of modernity with the civilizations of Greece, Rome, and Egypt. The imperative force of progress did not flourish in these civilizations for at least four reasons. There was no systematic investigation and manipulation of nature. Hence, they could not bend nature to fit our beliefs. Moreover, the concepts of "happiness" and "democracy" did not apply to all people, especially not the slave or the artisan, on whose shoulders much of the work for those grand civilizations rested.

Life was subject, furthermore, to an explicit yearning to move beyond the world of our senses and of our passions. This is especially true in Greek and Egyptian society. I will focus here on Greek society. Work and the development of the natural world was not viewed in a positive light by many Greeks because this world was shunned for the ethereal world of rational philosophical forms. Consequently, artisans and slaves occupied themselves with "inferior" technical activities, such as the building trades, while philosophers aspired to the nether world of ideals. Mechanical arts like engineering were called "adulterine arts." People of inferior (as the root word *adultery* suggests) heritage occupied themselves with these arts. Reason was the supreme

human activity. Consequently it aspired to the morally and intellectually lofty world of forms. Forms were thought to be the essential, rational nature of things. (This upward aspiration is somewhat less true
in Aristotle's thought.) Only those highly trained in thinking could
reach these sublime standards; no commoner polluting himself with
technical work could attain the highest realms. Hence, technical work
was inferior to rational speculation. Consequently, no universal imperative for progress was manifest in Platonic, neo-Platonic, Stoic, or
Aristotelian thought.

The notion of *moria* or fate, moreover, also inhibited a view of pro-
gressing reality. Moria represented a more or less fixed, unchangeable
order for life. Events, especially the future, were beyond the control of
humans. Even the gods seemed prone to the dictates of fate. The imperative of moria called for resignation not revolution, reformation,
or betterment. Any attempt at resisting fate would prove fatal. Thus,
individual artistic and technical achievements such as the Parthenon
were developed but they remained technical facts—isolated feats
amidst a more pervasive view of fatalism and stagnancy. Technological optimism would never have blossomed in this soil.

The Judeo-Christian tradition became one key step toward a more
progressive or unfolding, developing view of reality. A progressive
view means that new possibilities could be realized. Accordingly, the
Creation—nature and culture—forms the matrix for God's work and
purposes in history and is therefore intrinsically valuable. Development of nature into culture, the "cultural mandate," is associated
with the core of the Judeo-Christian tradition. This Hebraic mentally
undercuts, in principle, a Greek deprication of nature by viewing all
creation "very good." Further, it considers development a necessary
and essentially good part of the human activity, as long as "sub-
duing" also entails "keeping" or maintaining and preserving the creation, or all that is. Progress, in these traditions, has a more holistic
meaning. Holistic means that all of reality, in its many aspects, must
harmoniously develop if we are to say that the human condition is
improving. Note should be made of these notions of wholeness and
harmony; they will inform this entire work.

To be sure, medieval Christianity synthesized the core of Christianity with the thought of Aristotle and, therefore, encouraged an ambivalent attitude toward culture and especially technological
development. On the one hand, medieval monks kept learning alive,
promoted agricultural improvement, helped overcome slavery
through the development of the manorial system, and gave dignity to

work through the elevating of labor. Culture progressed, though with repeated and prolonged setbacks.

On the other hand, improvement of this world was not the final or most important goal of medieval Christendom. Heaven was the final goal, and this fact determined social priorities. The church and its edifices occupied the central place in medieval life and therefore attracted a great deal of artisan effort. The vast amounts of human and natural resources it took to develop medieval cathedrals give ample evidence to the centrality of the church for medieval life. Yet, if we look more closely at the construction of medieval cathedrals, we see evidence for ambivalence. Spires point upward to heaven, our final goal. Biblical themes, not secular themes, are etched in the stained glass windows. The cathedral often was constructed in the form of a cross, the central symbol of Christianity. The church became the hub for community life and hence dominated life. Medieval secular life was supposed to serve the sacred or church agenda, all of which, in turn, pointed beyond this Earth to heaven. There existed, therefore, an ambivalence about social progress in the Middle Ages.[14]

Modern technological and scientific optimism draws a portion of its spirit from the Renaissance—the "rebirth" of classical learning produced a hopeful air. This hope was based on a resurrected (from Greek thought) belief in the value, dignity, and the potential of individuals for social achievements, among them economic and technical accomplishments. Indeed, the origin of modern mass production arguably can be located in foundries of sixteenth-century Milan. By 1540 Vanoccio Biringuccio wrote *Pirotechnia*, or *Work on Fire:* this work became one of the most articulate statements on mass production written in that period. It was done in the vernacular so a wide variety of craftsmen could understand his more technical arguments. Improvements in metal crafting, along with the channeling of water into more productive efforts in the mines and foundries, simultaneously characterized the era's technology and created some of the fertile soil for optimism. The foundries of Milan began to chase away the demon of scarcity—so optimists thought.[15]

However, it was not until the Copernican revolution that the notion of progress and the related ideas of autonomy and Reason came to prominence. People freed their minds from the shackles of the church's external authority and past traditions, and shifted their attention to a hopeful future built by science and technology. Lord Chancellor Francis Bacon (1561–1626) could dream about empires ruled by science and technology. With a boundless enthusiasm and optimistic hopes, Western civilization set sail on the ecumenical—literally, whole-earth-uniting—ship of technological progress. Modern life resulted.

Sometimes this lust for freedom could not be obtained without violence. Freedom had to be won, by revolution if necessary, if progress was to be sustained. The preeminent symbol of this lust for freedom and consequent violence is the French Revolution of 1789. The French Encyclopediasts added another article of faith to the creed of progress before and after they overthrew the Bourbon monarchy (1589–1793). They believed in the perfectibility of the human intellect, morals, and, perhaps, even human bodies. Now the package was complete: mind, man, and machine could be tuned to perfection. Authority, superstition, heteronomy, religious faith, gender, and calendar—all external restraints— could be swept away and thereby progress was ensured. Educators touting the generative possibilities of rational intelligence and reformists working through social bureaucracies hoped to use social techniques to create a new world order after the revolution.

The Encyclopediasts also attempted to define human happiness in materialistic terms, a fact important in the understanding of Adam Smith and the rise of modern **capitalism**. Economic liberty and consequent material abundance were joined with progress to ensure the happiness of humankind, according to the Encyclopediasts.

It is important to note the cauldron of events occurring during the end of the eighteenth century. The Enlightenment gathers strength from the French and American revolutions and thus gives force to the dictates of Reason as well as to the quest for freedom. Further, Reason increasingly treats nature as a scientific object to be counted, divided, measured, and atomized. The goal was control. This control was gained through the Scientific Revolution that occurred at the end of the eighteenth century. The reality of the Industrial Revolution only confirmed and enhanced Reason's desire for freedom. The twentieth century certainly cements this drive to manipulate nature and is rewarded by mass production fed by an increased "plasticity of the natural milieu" (to use Jacques Ellul's poignant phrase). A pliable nature yields economic benefits: the chief preoccupation of American society. Could utopia be far away?

Optimism becomes a global, ecumenical outlook for members of the twentieth century. Thus: "Studies conducted by Soviet specialists and experts from various international organizations show that nuclear energy is now the only [*sic*] reliable type of energy that can satisfy our energy-hungry world."[16] This remarkable quote, one that so clearly evidences an optimistic spirit, betrays the confidence we speak of in this chapter.

Is the optimism born of fact or hubris? This question will be debated at the end of this chapter, as well as throughout this book. For now, let it be noted that the above quote appeared in a journal published two months before the Chernobyl disaster![17]

The erstwhile "evil empire" is not the only nation gripped by optimism. The *Challenger* tragedy challenged the pride, resolve, and technical and scientific expertise of the top NASA officials. NASA's overconfidence in their technical abilities led to a flippant attitude. Concerns about a faulty O-ring construction were dismissed. Dr. Richard P. Feynman, a world-renowned physicist and member of the President's Commission on the *Challenger* tragedy, declared that "NASA's (and Morton-Thiokol) managers had exaggerated the shuttle's reliability 'to the point of fantasy'."[18] Thus, it seems optimism is a spirit that transcends nationalities, socioeconomic ideologies, and geographic distances.

Julian Simon: A Modern Optimist

Modern optimists take up a question that demands an answer. What does it mean to be human? One optimist defines our central identity as that of the "ultimate resource."[19] We are capable of overcoming all humanity's thorniest problems, so says enthusiast Julian Simon, because we are "the ultimate resource."

Simon's work is one the best examples of a notion of linear progress, the core idea of historic technological optimism. Accordingly, improvements in technological areas must automatically translate into social and human betterment over time. This view holds that persistent problems such as economic scarcity, food shortages, land abuse, natural resource depletion, the corrosive effects of pollution, and the pathological effects of population density can all be addressed and their negative social consequences reversed into positive benefits. This optimism is sparked by Simon's belief in the unlimited potential of human genius and resourcefulness—hence, man as the ultimate resource.

Simon believes that there is no meaningful limit to the ultimate resourcefulness of humanity. Creativity forever will respond favorably to any shortage or human problem. To be sure, a short-run period of adjustment where genuine problems exist often occurs. For example, overpopulation is a problem in the short run because of the resources commanded to supply increasing numbers of people. Nevertheless, in the long run more people suggests more human potential to solve problems.

Is the globe facing long-term economic scarcity of resources? No, says Simon. Data indicates that natural resources have and will become *less* scarce in the distant future. Are we in the United States experiencing pollution problems? Yes, we have a current disposal problem, but we live, on average, in a less dirty and more healthy environment now than centuries before. We see on television the plight of famine victims and conclude that many parts of the globe suffer from food shortages. Simon argues that this idea results from a superficial look at the data. Per capita food availability has been improving for the last three decades. Further, famine increasingly has diminished for the last century. He expects such trends to continue indefinitely.[20]

Further, useable agricultural land is not diminishing. Because yields per acre continue to climb, the number of acres under cultivation have dropped. These and other noncultivated acres are being used for human recreation and the natural habitats for wildlife. Therefore, according to Simon, a better quality of life results for many humans and animals.

Simon argues on: natural resources are not shrinking. Natural resources will become progressively less scarce, substitutes increasingly more available, and the per unit cost for securing resources will fall in the years to come. This bright future also applies to energy use as a near "inexhaustible" store of cheap energy awaits our use.

To sum up, Simon believes that our future is bright because our standard of living is on the rise. Benefits will multiply in the future because productivity increases, due to population increases, will produce a longer, happier life for people. This happier life arises because there are more "ultimate resources," or people. Thus, "As I studied the economics of population and worked my way to the view I now hold—that population growth, along with lengthening of human life, is a moral and material triumph—my outlook for myself, for my family, and for the future of humanity became increasingly more optimistic."[21]

To what does Simon owe this optimistic sentiment? He locates his optimism in our constructive use of technology. Technology will produce more at lower costs and therefore will increase the quality of our lives; we have more for less. Technology also increases the amount of resources available for production and consumption. The pattern of an ever-expanding resource base at lower unit costs suggests, for Simon the optimist, a nearly infinite pool of resources. There will always be enough market incentive, technical supply, and consumer demand, as well as scientific know-how, to suggest a brighter future. We can expect economic prosperity in particular and progress in general to continue into the indefinite future.

> The fall in the cost of natural resources decade after decade, and century after century, should shake us free from the idea that scarcity MUST increase sometime. Instead, it should point us toward trying to understand the way that technological changes are induced by the demand for the resources and for the services they provide, and the way that such changes reduce scarcity in the past.[22]

Simon attempts to fortify his argument by quoting technological futurist and optimist Herman Kahn. Kahn focuses on the use of base metals, the foundation for production. He finds no limit to the store of metals available for future use. Indeed, all stores are either "clearly inexhaustible" or "probably inexhaustible." There are, therefore, no metals that are needed for production that will be depleted in the near future.[23]

HISTORY AS SIMON MIGHT SEE IT

To return to history as seen through the eyes of Julian Simon: The themes of linear progress, limitless economic possibilities, and the ultimate resourcefulness of humanity should be expected. These themes become the basis for his optimism. As humans marched on the evolutionary path, it was our technologies—stones fashioned into tools—that helped separate us from the animal kingdom. As we thereby progressed over millennia, our abilities to fashion metals into weapons and tools, to domesticate animals, and to husband the land helped us end our nomadic existence and therefore laid the foundation for civilization. While no systematic view of progress can be discerned in antiquity, magnificent technical projects nonetheless were erected. The Egyptian pyramids remain a symbol of ancient ingenuity. Carefully quarried stone, greased logs for the land transport of the rocks, substantial barges for water transport, skilled masons, and a basic knowledge of mathematics, as well as a slave class for raw power, helped the Egyptians engineer wonders of the ancient world. This feat represents the kind of ascent that continues.

If "all roads lead to Rome," then surely those ancient Roman roads must be sturdy, for they carried great weight and lasted hundreds of years. The roads built by Rome were durable because of the degree of engineering sophistication. Cut several feet thick, founded upon several layers of rock and gravel, and pitched at an angle with a tile and clay base so water could run off, these marvels of construction became the means of Roman conquest and subsequent bureaucratic control. Only this degree of engineering genius could sustain the weight of wealth brought to Rome.

It is a mistake to view the time after the fall of Rome as the "Dark Ages," if by that one means people were technologically ignorant. Technical progress continued during the Middle Ages, though with some ambivalence on the part of the pious.

It was the invention of the printing press that signaled the beginning of the end of the Middle Ages and the introduction of the Renaissance. The advent of moveable type, more malleable metals, and improvements in ink and paper helped set the stage for wider rates of literacy, a mandatory element for human resourcefulness. Now ignorance could be addressed by technology. The press and its accompanying paper were called the "great cosmopolitan and international instrument"—the printed word allowed release from provincialism and thereby promoted a literal worldview.[24]

The telescope expanded and even overturned humanity's view of the universe for all time. The restrictive views of the church could be overthrown for the more enlightened views of science. It was the physics of Galileo that was to lead to the first computers in the second half of the eighteenth century.[25] Remember that calculation was one necessary ingredients used in the domination of nature. The fruit of Galileo's work was to become primitive computers or externalized rationality, another tool in our arsenal of domination.

The Industrial Revolution, beginning in the early nineteenth century, gave birth to the use of large-scale machinery that became a symbol for an age. The speed, efficiency, power, discipline, and productivity of the machine became virtues, copied all through the "industrialized" world. According to Simon, the copious products, manufactured at lower per-unit costs, give a realistic air to predictions of falling per-unit costs. The mechanization of labor, or modeling human labor after the movements of the machine by

Frederick Taylor,[26] further enhanced productive capabilities. The machine offered laborer and industrialist alike the chance for a new birth or new opportunity for wealth and happiness.

The advent of the twentieth century witnessed the dramatic increase in food supplies. Simon the optimist must note the reality of famine; he is not a blind optimist. However, science did increase yield output per acre through improvements in plant breeding. Mechanized farm tools did increase the efficiency of agricultural production enough to meet rises in aggregate population, argues Simon. Thus, technology and science have and will improve indefinitely the plight of the hungry. This fact was true in the past, is true now, and will be true in the future, argues the optimist.

The twentieth century also witnessed the advent of mass electrification, perhaps the primary symbol of the century. Hydroelectric power and the splitting of the atom, along with the promise of various facets of solar power, have put at our literal fingertips a virtually limitless source of power. With little or no long-range problems with the production and consumption of water, uranium, and sun-related technologies, optimists like Simon believe cheap, abundant, clean sources of power always await us. Power to transport cold and warm air, to enlighten (via electronic books), to heal (via lasers), and to educate (via computers) await the creative spirit.

In summarizing, one must conclude that a bright future with a standard of living that matches and exceeds in quality our age awaits us in the future. The rising tide of people signals a greater stock of the most important ingredient—human knowledge—necessary for a better life, according to Simon. The promise of a better life includes a greater availability of natural resources at a lower per unit price, a cleaner environment, and lower energy costs. Indeed, there are no lasting limits to the upward more progressive trend in life. This upward trend is due to the "Ultimate Resource": *Homo sapiens*, man and woman, the wise, creative, knowledgeable giver of abundant life.

This excursus into an outline of the history of technology, as seen by the optimist Simon, was necessary to highlight a crucial equation that lies at the heart of technological optimism: increasing technical and scientific development yields greater amounts of economic well-being. This fact, in turn, is believed to raise the *overall quality* of life. Life, in its diverse aspects, is founded on and bettered in toto by complying with the technological imperative. This equation is known as linear progress.

Furthermore, most optimists understand the negative effects of technology; they are not naive. What they do steadfastly believe, however, is that human creativity and technical know-how can erase even the problems caused by technology. Alternative clean energy sources can be found to replace polluting wood burning. "Scrubbers" or smokestack cleaners can be installed to reduce industrial emissions and thus curtail pollution. Science can be applied to agriculture to increase grain yields while not eroding the soil. Refrigeration can be applied to reduce waste, and thus reduce pollution, therefore potentially increasing available calories for consumption. The list could and does go on for the optimist. The point: any technical or nontechnical problem can be solved by technical know-how. Optimists are thus strengthened in their beliefs.

Consequently, to human autonomy, Reason, and the general secular imperative of technical progress, we add the emerging force of linear progress. Linear progress fixes our hope for a better future to a projected historical outcome. Linear progress is not an ethereal imperative. Its march can be counted and plotted, and its assent fixed to specific technical inventions. Graphs thus increase our belief in technology because they give evidence of improvements in technical and scientific know-how. Figure 1 shows the progress of agriculture measured by kilogram grain yield per hectare from the mid-sixteenth century to approximately 2,000 A.D. and thus represents one concrete example of the beliefs outlined.

At least three points need to be noted. First, graph (a) is presented as accomplished grain yield. This graph gives clear evidence of upward or uninterrupted linear progress. In reality, graph (b) is more historically accurate in reporting the data that actually exists. The broken lines represent insufficient availability of data and interrupted phases of "progress." Graph (a) represents an interpretation of the growth of yield per acre that is insufficiently supported by empirical evidence. Further, it is based upon the hidden assumption or nonempirically tested belief that grain yield inevitably and in all circumstances moves upward. Thus, an interpretation and a hidden assumption and not empirical evidence effects the literal make-up of speculative graph (a.)

Evidence coupled with interpretation and hidden assumptions thus are determinative for the optimist's view of the future. Equally relevant for a view of the future is what is omitted from the graph. The long-term effects of heavy mechanization, chemical fertilizers, pesticides, and new crop cultivars are not known at this time. There is

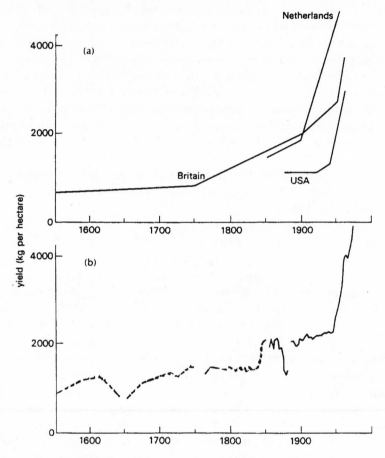

key to lower diagram
-------- impressionistic interpretation of scattered figures and known trends,
  for England only
———— government statistics for Britain (from 1884) and estimates by Lawes
  and Gilbert (1853–76), plotted as 5-year moving averages

some data that suggests that mechanized agriculture is detrimental to long-term soil stability. Thus, we may be causing long-term partial regression at least in soil quality because of the way we are doing agriculture. Again, actual future data may be much more uneven than the optimist suggests.

  Last, and perhaps most important to an understanding of optimism, is how belief functions in optimism. This foundational notion of belief effects data interpretation and consequently theory construction. Hope, confidence, trust in the inevitability of progress represents a core commitment of optimism. We hope and trust grain yields will continue to rise into the future and act accordingly. The obvious fact that the future is not available for us yet to manipulate is not part of the optimist's operative assumptions. The conditions under which we live are thought to be constant because the optimists believe they are created by the genius of science and technology. The constancy of human resourcefulness is thought to produce constantly improving conditions. Therefore, there is an empirically blind trust in the ability of technology to reap increased benefits for the future.

Moreover, there is another blind spot. Only willing to see the good of technology, optimists never do justice to the downside or negative effects of technology. Author Arnold Pacey rightly recognizes, therefore, this fact by labeling the points under discussion as "Beliefs about Progress."[27]

KARL MARX THE OPTIMIST?

Optimism has had a dramatic influence on modern global political realities. Karl Marx (1818–1883) was an optimist, and he has been one of the most influential figures of any century. Marx believed that people at their depth could be defined as *Homo faber*—Man the Toolmaker. In so defining humans, Marx broke with G. W. F. Hegel's rationalistic idealism that saw the mind as the universal force in history. Life is not driven by reason, said Marx. Rather, it is driven by the need to survive. Technology is employed to wrestle rewards from natural reality. The rewards for dehumanizing labor are economic in nature. The production and consumption of economic resources represent the economic dynamics of any society. Economic dynamics, of course, are the foundation for the march of all history, according to Marx.

The tools that are used to construct any productive system are the "forces of production." The forces are used to manipulate material reality. Material circumstances include raw metals, rivers, fields, air, and timber. They are literally at hand, ready for us to transform for our purposes. The material world forms the raw material for that mega-force Marx called culture.

In the Marxist view, modern civilization is gripped by the vile force called capitalism. Capitalism emerged from a revolt against the contradictions and the limitations of the guild and manorial serf system of the feudal age. It has provided the impetus for economic and technological growth—up to this point. Marx thought that capitalism's time of influence was about to end because of the monopolies of wealth and power created by the capitalist ruling class. Because the ruling class would not surrender their political and economic power lightly, a revolution was needed. This revolution would occur when workers became aware that their labor was being purchased at a fraction of its real worth and, thus, they were being robbed. Factory owners sold labor's product at the highest price the market and their power would allow. Profit was, consequently, the excess money remaining after subsistence wages were paid to the laborer and capital was invented, modernized, and enlarged. This process inevitably led to the concentration and control of capital, according to Marx.

The concentration of wealth is predicated in part, therefore, on the expanding technological basis of modern production. The workers were vulnerable in this scenario. The multiplication of machinery results in the replacement of man by machine. The replacement of men by machines is called structural unemployment. This technical unemployment, the exploitative labor conditions, and subsistence wages forces immense worker discontent.

Universal worker revolt was predicted by Marx. However, the coming revolution does not diminish the importance of technology for social betterment. After the revolution, the workers were supposed to gain control of technology and garner its fruits for the good of their entire life, increasing both its duration and quality. Improvement in personal and social relationships caused by a more equitable distribution of technology were believed to follow the revolution. More philosophically, that the forces of production and the relations of production are harmonized in the triumph of the formerly alienated worker. The newly enfranchised worker reappropriated the surplus value for his own materialistic ends. This utopian state of bliss needs technology so that a postcapitalist world can maintain the conditions of material abundance. Since the inevitability of material abundance is guaranteed, Marx could say with aplomb, "to each according to his means, to each according to his ability to contribute."

> Thus . . . the modern technological phenomenon acts not only as a goad to profoundly needed changes in society, that is, a stimulus for revolution . . . but also as a lure to a new equilibrium of productive forces in social arrangements that will fulfill all human hopes.[28]

Marx's attitude toward technology finally is **utopian**. Indeed, Marx believed that "a revolution in the method of production in one sphere of industry involves a similar revolutionary change in every sphere of life."[29]

R. BUCKMINSTER FULLER

Surely, thinks the perceptive reader, a technological optimist cannot be found in the twentieth century, with its plethora of technological destruction. However, there is the example of R. Buckminster Fuller (1895–1983). His work demonstrates a religiously optimistic thrust. His is an interdisciplinary analysis of the place and importance of modern technology. Fuller's gift for understanding philosophy, "religion" (traditionally understood), mathematics, literature, engineering, economics, and politics enabled him to develop a worldview in which technology generally, and engineering specifically, played a dominant role.

Fuller is perhaps best known for his design and construction of the geodesic dome. The dome combined the properties of the tetrahedron, or triangular pyramid, with those of a sphere. Increasingly strength and reduced weight per unit were key features of this structure. A relatively lightweight, smallish dome was capable of providing great interior space and sufficient strength. This space could be more efficiently used, thereby reducing costs while providing natural light and more efficient function.

Fuller was a first-rate engineer whose talents extended beyond the technical to the humanistic disciplines. Fuller was conversant enough with philosophy, for example, to flavor his technical writings with philosophical insights. Reason, so Fuller believed, dominates history. Fuller is one of the best contemporary examples of this belief in Reason. A priori, deductive reason characterizes humanity. This reason gives humans the ability to reflect critically on and consciously transcend daily experience. This faculty enables us to be critically aware of ourselves and our natural and social environment. Practical or engineering intelligence emerges out of this reflective, self-aware posture.

Practical rationality is instrumental in character. Instrumental rationality develops technical means to the end of control and problem-solving. More philosophically, Reason's pragmatic job is to develop instruments that successfully lead to a solution for a recognized problem. The value of ideas, therefore, is in the mastery it gives humans over a defined situation. Mastery signals progress.

The following poem demonstrates Fuller's view of instrumentalism and his view of "God," or his ultimate commitment:

> Though you have been out in
> a froth-spitting squall
> on Long Island Sound or
> in an ocean liner on a burgeoning sea
> you have but a childlike hint of
> what a nineteen-year-old's reaction is
> to the pitch black shrieking dark out there
> in the very cold northern elements
> of unloosening spring
> off Norway's coast
> tonight
> 15,000 feet up, or
> fifty under or
> worse,
> in the smashing face of it and
> here I see God.
>
> I see God in
> the instruments and the mechanisms that
> work reliably,
> more reliably than the limited sensory departments of
> the human mechanism.
>
> And he who is befuddled by self or
> by habit,
> by what others say,
> by fear, by sheer chaos of unbelief in
> God
> and in God's fundamental orderliness
> ticking along on those dials
> will perish
> and he who unerringly
> interprets those dials
> will come through.[30]

Fuller muses in this poem, written about the outbreak of fighting near Oslo, Norway, during World War II, about the nature of God, instrumental Reason, and trust. Whether or not the term "God" refers to a theistic deity is open to speculation; Fuller does not seem to answer this question. This poem manifests an ultimate conviction that instrumental Reason can be trusted above all. Reason makes the technical instruments that mirror a God who through these instruments delivers us from defined problems. Our physical senses will certainly deceive us while instrumental Reason will not.

Fuller must order the universe because it is disordered, prone to chaos, and in need of a plan. The future holds only randomness and disorderliness, according to Fuller. This chaos and randomness will continue only in the absence of the active reordering of reality by human Reason. Humanity brings a new fundamental reality that results from the universal law or ordering principle created by human Reason. We bring on "omnicon-tracking, convergent, progressive orderliness . . . to the universe."[31] Reason is reordering reality, thereby promoting harmony, peace, and progress. Reason is Fuller's ultimate force, and it is to Reason that he pays ultimate allegiance.

Engineering is the practical profession for realizing his rational ideals in culture. Engineers implement what a priori and instrumental reason[32] demands of us. They incarnate, literally, as we will see in a moment, the "word," into cultural existence. Furthermore, engineers transform nightly reality, thereby giving it order and purpose. The transformation of so many chaotic elements by a machine tool makes humanity the master, the conqueror of reality. The wisest, most courageous master of all is, of course, the engineer.

I close this section on Fuller with one more fitting credo to humanity and to technical rationality so described in this chapter:

> . . . tonight vividly (as tacitly always)
> God is articulating
> through his universally reliable laws.
> Laws pigeonholed by all of us
> under topics starkly "scientific"—
> behavior laws graphically maintained in the performance
> of impersonal instruments and mechanics
> pulsing in super sensorial frequencies
> which may serve yellow, black,
> red, white, or pink
> with equal fidelity.
> And I see conscious man alone
> as mechanically fallible
> and progressively less reliable
> in personal articulation
> of God's ever swifter word,
> which was indeed in the beginning.

Only as mind-over-matterist,
as philosopher, scientist,
and informed technician
impersonally and universally preoccupied
is man infallible.[33]

TECHNOLOGICAL FUTURISTS

A final word about technological futurists: I have shown how graphs on grain yields can be affected by interpretations built on the hope promised by progress. Technological futurists do just this. They take existing data and project current trends into the future optimistically expecting a good outcome. Typical of this approach and the underlying optimism are futurists Alvin Toffler and John Naisbitt. Historian Howard Segal surveys the works of these popular futurists. Segal's work portrays his pessimistic attitude toward modern technology. Thus, it serves as a good transition between our explanatory and evaluative stages. Essentially, Segal maintains that optimists approach the future with a basically hopeful attitude, tempered at times with pessimism, because they conveniently forget the past and its lessons, which show that touted technologies manifest a great many more difficulties than enthusiasts once thought. Future technological projects are promoted vigorously by large corporations as a cure for this, that, or the next traditional human problem. Allegedly, the market then disperses the cure *because* we demand new technologies. This is part of the hype, not reality.

Technology has not solved abiding human problems, nor has it been without its own difficulties. Segal rightly concludes in response to Toffler and Naisbitt that optimism is supported by "high-tech prophets" whose uncritical "faith" supports future utopian visions of technology. These high priests of technology are dogged by deep ironies and contradictions, however. Chief among these contradictions are the elements of tragedy, mistake, hubris, and social and human regress. Perhaps this is the deepest irony of all: Technology has brought with it regress at the very moment progress was expected.[34]

THE PLACE OF TECHNOLOGY

Optimists believe that technology should occupy the greatest possible space and should be accorded the highest level of importance. If technology is the key to solving lasting human dilemmas, and if humans can be defined fundamentally as technical creatures, then its place should grow because technical activity is essential both for human identity and for deliverance from perpetual human problems. The consequence of this view is a cultural environment increasingly defined by technology. We become inundated by modern technology because solutions to deep problems are addressed by predominantly technical means. Thus, optimists' central assumptions and inner rationale inevitably lead to the expanding ontological[35] place afforded technology. Technology must expand given the optimists' worldview. There is an inevitability to the increasing place accorded technology. It is inevitable because the optimists' deepest convictions naturally lead them to expand the place of technology. These deepest convictions, such as belief in the inevitability of linear progress, thus influence subsequent actions.

Optimism contributes to the constant expansion of technology for other reasons, as well. If technical reason is believed to constitute the universe, as with Fuller, then the place accorded to technology will grow. If our technical nature matches that of the universe, then the essential nature of the universe mimics our nature. We control parts of the universe; why can we not control ourselves as well? This control is attempted for at least two reasons. First, because reality mirrors our nature, we are familiar with the essential technical nature of reality. Hence, we can manipulate it because we have experience of the technical nature of reality through self-knowledge. Second, human facilities such as Reason, physical traits such as opposing digits on our hands, and even the character of "a god," are evolutionarily suited to technical control and manipulation. Hence, technology expands.

Further, if technology is central to human nature, then the influence of technology will spread with the expansion of the human species. When we are technological we are becoming more fully human as we conquer technical reality. Moreover, if there is success in one extratechnical area, such as the economic aspect of commodity production, then other areas of life should be amenable to the same kind of success through the same kind of technical means. Thus, Frederich Taylor changed the nature of the workplace by persuading the worker to pattern his movements after that of the machine.

This logic leads to the conclusion that humans must become mere technical-instrumental means not ends for two reasons. First, in spite of all of one's humanness, we must become someone's technical means used for another's technical ends. One's essential technical and instrumental nature demands that each person be treated as a means. Does not the traditional designation of labor as a "factor of production" take for granted that human laborers are mere means? Thus, in our increasingly technical society people complain of a loss of meaning, a feeling of insignificance, and being the brunt of another's agenda. Could these sentiments be caused by the increasingly technological nature of our lives?

It is easy to see why the optimists have succeeded in expanding the import and the place of technology in our lives. Technology is believed to solve enduring human problems like hunger, develop our essential human nature, and mirror the essential nature of the universe. Technology, given this reasoning, becomes an ethical imperative because it is thought to fulfill all that is necessary to save and sustain life. The expanding place afforded technology is thought to be necessary because the rewards greatly exceed temporary costs. Should technology create problems, then technology could remedy even these problems. Hence, the place and import of modern technology grows.

Does technological optimism encourage an overabundance of technology? Do optimists through their belief in the salvific effects of technology end up erecting a technology that emasculates other areas of life? It is to this and related questions that we now turn.

Optimism Discerned

The reader may be tempted to launch into an immediate critique of this seemingly naive position. Indeed, this position does seem to lead us to a "technicism," or to the exaggeration of the place and the importance of technology. Are the optimists led to this exaggeration because they view reality as primarily technical/mechanical and instrumental? This evaluation will attempt to discern strengths and weaknesses of optimism using key ethical principles.

The development of technology has helped improve the quality of human life. This general evaluation results from the reality that humans are, by their nature, at least partly technical. Hence, developing our technical abilities means developing at least part of our human identity. Developing nature, securing technical progress, reducing some human want, ignorance, and fear have been among the achievements resulting from the development of technology. This fact has increased human dignity because part of our essential nature has evolved to meet our needs. Therefore, practicalities like a reduction in the burden of labor, the impact of dreaded diseases, and our exposure to the natural elements represent the practical result of the quest for human dignity.

When the human spirit is freed from unnecessary restriction, such as happened during the Renaissance when the institutional church lost some of its dominance, then creativity and resurgence grown as expected. While the term "rebirth" may be overstated, the human condition was advanced by the Renaissance's development of life beyond the church. This is so because the principle of a diverse reality means that life in its many aspects must be more or less simultaneously developed. To restrict life to one or a few aspects, as medievalism tended to encourage life to be dominated by the church, narrows life's possibilities.

Human dignity can and must be expressed in the harnessing of nature for human survival and development; this the Enlightenment and the Reformation did. Humanity is the only species capable of directing the potential of all of nature to some end. Humans inevitably will be technological; it is a necessary and legitimate human activity. When technology addresses noteworthy human problems, and to some extent alleviates the burdens posed by the problems, then the dignity of both the receiver and the operator of technology is enhanced. Wisely developing the earth's potentials by technical practices leads to the enhancement of human dignity because it invigorates our sense of creativity.

Nature has been harnessed to a degree. Ores have been made into steel that strengthens our structures of transport and protection. Timber has been cut and fit to cover our shelters, thereby enabling us to build houses. The wind has been used to *propel* technical objects like ships and windmills. Air has been harnessed to propel cars (via the carburetor) and airplanes. Fire has been tapped for a plethora of purposes. Examples could be multiplied, but my point stands: the skillful forming and developing of nature, especially after the scientific revolution of the seventeenth century, increasingly leads us to enhance our life by addressing our universal needs. One valuable human activity is being technological. This is **normative** behavior in the sense that humans, in so doing, are fulfilling one of their created and therefore necessary human functions.

Further, because life is whole, technology not only affects the rest of life, but the rest of life impinges on technology. An example of the inherent ethical force of technology is appropriate. This example shows how nontechnological factors may be contributing to problems.

The advent of mass production in the Industrial Revolution has solved, in principle, the age-old dilemma of want. Now we can produce enough goods to address adequately many human needs. While solving the problem of want, however, we have not begun to solve the problem of distribution. Ignorance could be addressed through mass-produced textbooks. The burden of labor has been eased by the advent of the modern machine. We are well-sheltered from the harm of the elements by the works of milled timber and mined ores. Modern agriculture produces enough food to feed the world. Yet in all of these examples, factors such as enmity, greed, hatred, and fear can and often do prevent a more equitable distribution of food. To recognize this reality is to admit that factors other than technological ones must be addressed if technology is to enhance life. Optimism cannot admit these realities.

Finally, and most obviously, technical progress enhances human functions. Machines are to be preferred in many instances over animals or human muscle as a source of power. Machines certainly exceed our muscles in endurance and precision. The clock gives us greater precision in telling time than does the sundial. Modern science and medicine exceed the capacity of our eyes because of the invention of the microscope. Technical progress has contributed, therefore, to the betterment of the human condition. Development is an ethical principle because stagnancy is a waste of human potential.

Further, thinking in an orderly, calculated, systematic, probing manner results in a great many benefits. Gratitude should be expressed to scientists for the fruits of their methodology. It must be noted that there is nothing inherent in the methodology of modern science that prompts such an exaltation of reason. Our concern is not with the methodology of science per se, but with the veneration, exaggeration, and trust placed in a fallible human activity that we have called reason.

One may go further and say that instrumental rationality—using technical reason as a means to another end—does not trouble us. What troubles us is the believed absence of greater ends, values, commitments, beliefs, and implications present in and through instrumental rationality. The brilliant men who developed the Bomb were aware of the implications of their experiments and therefore must be held accountable for their results. There are no value-neutral engineering projects because designs are drawn up with integral values, principles, and goals in mind. When an engineer focuses on technical/instrumental rational practice, thereby forgetting other human needs, then a myopia occurs. This over-concentration on technical rationality as means with only technical ends in mind, too prevalent in today's engineering programs, represents a truncated technical worldview, and will lead to an erosion of human values.

As I have said, optimists are not naive; they understand that technology can have negative side effects. Nevertheless, optimism believes that the "good," or benefit, of technology predominates over the "bad," or risks and negatives. Optimists would not deny that a given technical object can do harm. They simply believe that a harm can be remedied by another technology. Cars cause pollution, and catalytic converters reduce it; so says the optimist.

Because technical objects must be multiplied to remedy negative effects of past technologies, one might say that optimism leads to a saturation of our culture with technology. The optimist must create technological objects to remedy technological and nontechnological problems. This is not redundant grammar, it is redundant technology that begins to suggest technicism. The seeming self-multiplication and the resultant overextension of the place and the importance of technology are called technicism. Technicism represents the major concern we have with optimism.

At the heart of technological optimism lie some beliefs that must be questioned as well. Is the universe essentially rational/technical as optimists would have us believe? It seems that the "irrational"[36] is just as strong as is the rational. Storms, earthquakes, irregularities, birth defects, and human evil all seem to testify to the absurdities manifest in life. Rationalism takes one necessary human function and believes that human nature and the constitution of the universe revolves around that function. Why should rationality be the supreme factor for all of life?

The liberal arts curriculum testifies to the fact that there are many areas of life, all of which could claim to be central to human identity. When technical rationality is thought to be essential to all existence, then it is exalted. When that is the case, as with Fuller, then the exaggeration of Reason follows and we are led to believe that reality is essentially instrumental in character. Technicism, therefore, results from a faulty view of the person and of reality.

Even if I grant that technology can correct most human problems and that, as a result, its presence should be multiplied (which I do not), then doesn't this state of affairs create an increasing imposition of technology on modern life? In fact, isn't this imposition the key problem that results from the deployment of modern technology? If it is, then to grant technology more room to attempt solutions would be only to increase the problem caused by modern technology. Perhaps we need to take a few mental steps back and reconsider our fundamental assumptions.

Some questions follow this reconsideration. Does human technology participate in the human condition? (That is, is technology subject to the same kind of distortions manifest in all branches of human endeavor?) Can culture become so saturated with technology that the society may be called technological in nature? Can the problem that led to technicism be located in a lack of a sufficiently broad vision? Specifically, is a view of what it means to be human and a view of the nature of reality causing a myopia?

The answer to the first question seems to be apparent. If technology is a human project, then it participates in the best and the worst of our nature because it is inherently tied to our nature. Only a **positivistic** view, one stressing the artificial separation of technical facts or thinking from values or principles, could sustain the seemingly naive belief that technical objects can be separated from the human condition. A view that ties modern technology to our nature has to address an ironic point. Optimism has unleashed a force, which while tied to the human condition, inherently exceeds and therefore renders obsolete the mimicked human function. Thus, to the degree that technology saturates our lives, to that degree human functions and the human personality are threatened with obsolescence. The machine exceeds in use our muscles. What then shall we do with our muscles—invent more machines to exercise our muscles? Indeed, because the machine has reduced our need for muscles, our health is affected. This fact helps explain the modern penchant for machine-driven exercise.

If humans are capable of serious, even monumental errors like the Chernobyl and *Challenger* disasters, then our objects will follow in this path. The "fantasy," "exaggeration," and "overconfidence" that prompted these disasters are, legitimately, prompted by pride, arrogance, and ignorance. Optimists are loath to admit the presence of these traits in humanity. Consequently, limitations, errors, and technical malfunctions were ignored and, therefore, their consequences repressed. Thus, the dignity that comes with being technological has been perverted by arrogance and pride. Similarly, the legitimate harnessing of nature has degenerated into the concrete jungle, with its gridlocked animals we call cars: dominion has become damnation, as countless accident victims and our punctured ozone layer testify. Arrogance can lead to the destruction of life.

Further, the second question (again the relationship of technology to core human identity) suggests another problem. Optimists have cheapened life, perhaps unwittingly, by linking humanity's core function to instrumental or means-to-end rationality. This stance inevitably leads to treating humans as mere means to some greater technical ends because technologists act consistently when they treat people as technical means to greater ends. Given their belief about the technical nature of reality and the person, they can do little else. Optimists violate the wholeness of humanity because of their myopic equation of human nature with technical rationality.

I now arrive at an evaluation of the heart of technological optimism: the notion of linear progress. To repeat, the notion of linear progress equates technological betterment with improvement in the human condition over time. That is, improvements in one technical field at one or even several points of time are equated with the future betterment of the human condition. The graph on British agriculture and the surrounding claims by optimists make this clear.

However, improvements in agriculture depend on a context. The land, the labor, and the energy required for growth of this much grain is the context. While grain yields per hectare may increase, problems arising from the use of improved technologies and highly mechanized and fertilized land use may negatively impact, and often do, the context. Land erosion, labor strife, and soil depletion have been caused by improvements in technology. Chemical fertilizers and pesticides, mechanization, hybridization, and even genetic engineering temporarily may increase yield or correct specific problems. However, these techniques also can cause short and long-term *regression* in ecosphere quality and, therefore, threaten long-term yields. In short, technical progress can and often does cause a regression in other aspects of life. The relationship between green-house produced gases and global warming is a case in point. Optimists who hold to this notion of linear progress are often blind to this reality.

It is a **reductionism** on a grand scale to equate technical progress with human betterment. More important than technical triumphs are the needs of other areas of life—needs that transcend but are not unrelated to technology. Plants need air, people need to be protected from harmful rays, soil needs rest and rejuvenation, and animals need habitat. It is sheer fantasy to believe technology alone, or even fundamentally, provides for these needs. To indulge the facile equation of human betterment with technical progress, as those who hold to linear progress do, is to reduce the needs of reality down to a thimble-sized container. If one believes that reality has many equal needs that must be coordinated with technical ones, then a more holistic perspective about betterment is needed if we are to prosper humanly.

Furthermore, the idea of linear progress, as in Simon's work, carries with it an optimistic outlook for the future. The diagram on grain yields, showing increases from 1945 through the present and to the future, illustrates this point. Believing in the inevitability of progress for the future, optimists' projects are often oversold. No available data exists; hence, it is speculative to conclude at what levels we may use land, labor, and energy indefinitely. In fact, current major American efforts at soil reconstitution suggest otherwise. Thus, without solid data that can be gathered for a future date, neither the rate nor the fact of growth can be assured. Therefore, if one were simply to posit an untested, supra-empirical claim that grain yield will continue as per today's yield, one may legitimately conclude that belief, not fact, was present.[37] Indeed, this fierce and abiding trust in human reason should be admitted.

The heart of the problem with optimism is not to be found in pride, although as I have argued that is a problem. Rather, the heart of the problem is located in autonomy. Seeking to become a law unto ourselves, we have attempted to become autonomous from all heteronomous forces, especially God. In our desire to be autonomous, we think that we can win our freedom by the subjugation of nature. Thus, philosopher Herman Dooyewaard notes:

> Proudly conscious of his autonomy and freedom, modern man saw "nature" as an expansive arena for the explorations of his free personality, as a field of infinite possibilities in which the sovereignty of the human personality must be revealed by a complete *mastery* of the phenomena of nature.[38]

This attempt at mastery, which has brought great success, was accomplished not only through Reason in general but instrumental rationality in particular. Technical thought becomes an instrument or a means to the end of subjugation of nature. This subjugation was not only to shape nature to our whim. More importantly, instrumental rationality sought to reduce the meaning of nature to a mechanical process. This mechanistic outlook represents an extension of the root problem I am calling technicism. If reality is being remade by instrumental rationality, then reality becomes predominantly technical in nature. If it is technical in nature, then all traditional human problems must be amenable to technical solutions. Thus, life in its many activities and aspects becomes reduced to autonomous rationality.

I believe that this view of rationality and the autonomous spirit that informs it can be said to lead to an ideology of technology. Thus, noted philosopher of technology Egbert Schuurman writes:

> In short, technicism, or the implicit ideology of technology, is the dominant expression of the humanistic groundmotive. Technicism entails the pretension of the autonomous man to control the whole of reality: man as master seeks victory over the future. He is to have everything his way. He is to solve problems old and new, including problems caused by technicism, so as to guarantee an abundance of material progress.[39]

The notion of a "groundmotive" draws its validity from the notion of the depth of life. Life is rooted for entire communities, persons, and civilizations in something, someone, or some process. The collective grounding of entire peoples and cultures may be called a groundmotive.

The reward for the subjugation of nature is not simply amenable reality, however vague that may sound. Rather, the reward is tangible economic fruit. Economic rewards, such as products, salary, and consumption, represent key rewards for technical subjugation. Humanity has wrestled nature, and the winning is tasted in the bite of economic rewards. At this point, instrumental rationality is joined by utilitarian rationality. Acquisitive, calculating, economic rationality takes the specific outcomes produced by technology and labels them by a calculus known as utilitarian. Technology produces discreet and countable bits of economic rationality waiting to be enjoyed. Enjoying these bits results in "happiness." Happiness or utility comes from the consuming of economic units. We are believed by economists to be utility-maximizing individuals who seek to weigh, calculate, and amass bits of measurable happiness produced by technology.[40]

Nature and people associated with nature will not sit idly by while this subjugation takes place. In response to one exaggeration, another exaggeration of approximately equal intensity has arisen in Western civilization. This equal and opposite exaggeration poorly attempts a harmony or a compensation. I call this particular over-reaction to technicism "naturalism," or the idolatry of nature. One can see this exaggeration in the "ecocentrism" of the counterculture of the 1960s, the earth-goddess Gaia myth, and many current New Age movements. Perhaps the most influential current movement is called "deep environmentalism." In direct response to the technicism and the control and degradation of nature, naturalists seek alternative technologies rooted in nature's principles. Deep environmentalists believe the Earth provides the only rules and laws capable of sustaining life. Nature, however, is more than merely preserved; it is venerated in environmentalism.

Nature is seen as a large, interconnected biological organism that is believed in naturalism to be invaded by a foreign mechanical object in technology. Technology is foreign because the biological organism is natural and interconnected and therefore normative, even while technology is mechanical, fragmented, and inherently unethical.[41]

This naturalism and ecocentrism have popular expressions as well. Bumper stickers proclaim Earth as our "mother" or as a "spaceship," signifying thereby a view that Earth gave us birth (and should be venerated as we would our real mothers). This "spaceship" carries us on our wonderful journey called life. This reduction of the heart of life to nature and consequent exaggeration of organic processes represents a poor alternative to technicism. An exaggerated technological control is juxtaposed to an equally absolutized naturalistic freedom that believes that freedom can be found supremely in biological-organic nature. This tug of war between technical control and organic freedom represents the battle of two, more or less equally strong, titans making up what is called a **dialectic**. While these titans clash, however, no way can be found to transcend this dialectic or inherent contradiction: only new dialectics can be posed.

2
Technological Pessimism

INTRODUCTION

I have just shown that technological optimists have an abiding faith in technology's power to solve long-standing human dilemmas. This faith rests on Reason and its attempt to subjugate nature. The rewards for this labor are economic fruits called utilities. The perceived human betterment provided by technology is called progress that can be mathematically plotted and linear or sequential implications drawn. I have also shown how biological, human, and the Divinity's nature have come to be defined by a technological consciousness. "Naturalism," or a complete worldview occasioned with technicism, attempts to remedy a one-sided technicism by an equal and opposite one-sidedness. Thus, a dialectic, or a deep contradiction exists in the West, that being a contradiction between the necessity of technique and the freedom of nature. Technical rationality initiated this dialectic. This technical rationality concerns the pessimists.

This chapter will survey the thought of some representative technological pessimists and their negative views of the place and the meaning of modern technique. Typical of this negative view of the influence of technique is the thought of Jacques Ellul, and specifically his definition of **technique.**

> [T]echnique is the TOTALITY OF METHODS RATIONALLY ARRIVED AT AND HAVING ABSOLUTE EFFICIENCY (for a given state of development) in EVERY field of human activity. Its characteristics are new; the technique of the present has no common measure with that of the past.[1]

Technological pessimists have argued that the meaning and the imposition of modern technique is absolute, as the quote portrays. Its

influence must be avoided if life is to retain its rich character. There is
no facile optimism, no calculations of the trade-offs between the positives and the negatives of technique here, nor any attempt to locate
technique within the context of life's many demands.[2] Technological
pessimists see only the dark side of technique.

I have chosen to examine three representative, interdisciplinary figures—Jacques Ellul (in the main), Jürgen Habermas, and Nicholas
Berdyaev—for two basic reasons. First, they represent two important
intellectual traditions: Ellul and Berdyaev are Christians, and Habermas is a **neo-Marxist**. Second, their command of the breadth of
knowledge qualifies them to address so crucial a topic as the place of
technique in our lives. Certainly, there are other representative figures, but these three figures, in my opinion, uniquely confront the
modern reader with the problems of technology in disciplines as far-
ranging as theology to the philosophy of mathematics.

I will also note that the two Christian authors employ theological
categories in conjunction with their sociological and philosophical
analyses of technique. A recognition of this fact is mandatory for understanding. Our insight into the meaning of technology lessens and
becomes proportionally more superficial when any discipline is ex-
cluded. This task of understanding the whole of one's thought is especially important for the "dialectical" thought of Jacques Ellul.

JACQUES ELLUL AND PESSIMISM

One of the most widely recognized and respected technologists is
French social theorist Jacques Ellul. He has thrown down the gaunt-
let to the modern reader by arguing that technology has not produced
a heaven on Earth; it has spawned a gulag. The technological prison
that surrounds and defines us is totalitarian, autonomous, "demonic," and insidious in character. In fact, it is so all-encompassing
that society can be defined as technological: that is, society exists by,
for, and unto technology, according to Ellul.

Central to Ellul's understanding of technology's place in society is
his understanding of the social influence of technique. Using words
like "totality," "rationally arrived at," "absolute efficiency," and
"every field," the definition cited above conveys an absolute social
domination. That is to say, technique is a sociological fact that canni-
balizes social life. No area of life escapes its iron grip. We can find no
refuge in a fantasy world.

Autonomous rationality orchestrates this social process. This rationality is autonomous in the sense that it allegedly operates on its own laws. The goal of this rationality is the efficient control of every area of life. The methods—scientific, investigative, technical, productive—of technical experts are the avenues used by autonomous reason. The social influence of these technical methods is both extensive and intensive. Technique deeply transforms the fabric of life. People, the natural world, the workings of science, views about our humanity, traditional religion, as well as art and politics, all come under the tyranny of modern technique.

It would be a mistake to believe that Ellul equates technique with technical objects. While tools, weapons, cars, computers, and so on are technologically made, these objects do not represent the heart or core meaning of technique. Rather, technique is an autonomous social process. Technique produces a rational, step-by-step, procedural way of living. Technique can be practically seen in factories, bureaucracies, research and development teams, city planning, and methods of one kind or another.

The "expert" is the representative designated to demonstrate clearly and explicitly a technical mentality. Experts are "specialists" who "master" (or control) knowledge in a field because of their "competency" (or narrowed specialties) to evaluate and "solve" (by rational means) problems. Professional acceptance and self-esteem come with the credentialing process that labels one an "expert." These experts, in turn, impart methods or techniques to us for solving life's problems. We trust these experts to give us advice aimed at reducing problems in life. However, Ellul contends that in trusting these experts for solutions we lose our freedom and integrity.

TECHNIQUE'S ATTRIBUTES

Technique is first and foremost "religious" in two senses, according to Ellul. The word *autonomous* literally means that technology has become a law unto itself. "Technique has become autonomous; it has fashioned an omnivorous world that obeys its own laws and that has renounced all traditions."[3] Technique is also universal; its reign is so complete that traditional religions and associated mores fall prey to its sovereignty. The Western technical and economic secularization that significantly remade the Shah's Iran is one of the best examples of technique's ability to overrule traditional Islamic society. Customs and values, such as a male-dominated relatively rigid social system, and a hatred of Marxism and capitalism (in theory), were eroded by technique. In their place came the Western "virtues" of materialism, efficiency, modernization, production, and "equality."

Technique was not restricted to the Shah's Iran. The means used to overthrow the Shah and embarrass the United States were those of modern technique, as well. Were not manipulated television images, brandished weapons, petrodollars, and social techniques such as demonstrations among the various techniques used? Thus, technique is used to remake society more successfully than the Islamic faith.

There is a second way that technique is religious. Modern technology has usurped the place of the sacred. No technical deity is venerated because the deity is not seen and therefore is not amenable to manipulation and calculation. In place of the deity, we "praise" technology. While praise is not literally doxological, it is veneration nonetheless. The optimist's faith in humanity is projected onto the technical object constructed by instrumental rationality—this was discussed in chapter 1. The triumph of human ingenuity becomes a "technical triumph." The pride that wells up within us when, for example, the steam engine is unveiled is attributed to the object itself. Thus, Ellul contends,

> Nothing belongs any longer to the realm of the gods. . . . The individual who lives in the technological milieu knows very well that there is nothing spiritual anywhere. But man cannot live without the sacred. He therefore transfers his sense of the sacred to the very thing which has destroyed its former object: to technique itself.[4]

Technique is the primary and unique form of modern **desacralization**.

Technique's power is not restricted to the domain of the sacred. Technique's power was palpably manifest in the bureaucratic process that marshaled me through my graduate degree. Seemingly endless streams of divisions and subdivisions, administrators, rules, forms, and procedures constitute the modern form of authority for the large university. Efficiency, speed of bureaucratic access and exit, and expert knowledge became the key intellectual skills needed to navigate through the bureaucracy. "Specialization," mastery, and route recall were the intellectual skills needed to complete my education pipeline. This process extends beyond my graduate work to modern academia.

Ellul maintains, moreover, that technique dominates science. Methods, procedures, classification, controlling variables, and manipulation of material (such as gene manipulation) suggest that science itself is dominated by technique.

Certainly sport and play are not dominated by technique, objects the free spirit. Play represents the free movement of the human body according to an athletic imagination. While this may be true in childhood, consider how children are trained in play. The effervescent youngster is introduced to techniques that enhance winning. Specialization is encouraged at an early age. Motion, thought, and raw ability are transformed into efficient movements and well-practiced, repeatable operations that are machinelike in their quality. Is it any wonder, therefore, that we speak about our sports teams and use technology to dress up our playing fields? The Cincinnati Reds were called the "Big Red Machine" in the 1970s because of their power, efficiency, speed, and productive capabilities. These professionals often play on Astroturf, a space-age product the name and function of which serves to separate play technologically from nature.

Technique is no respecter of economic views or geographic boundaries. Industrialization is an example of a world-encompassing ideology.[5] Communism, socialism, and capitalism are all driven by the same mechanical techniques.[6] We know Marxism and socialism are infamous for their methodological control of the market. Is capitalism similarly controlled? Is the market free of the grip of technique?

John K. Galbraith argues in *The New Industrial State* that markets are no longer free, if by free we mean separated from large governmental, organizational, and bureaucratic control. The larger and more complex the technology used in competitive markets, the greater will be the requirements of specialization, capital commitment, and most importantly, market manipulation. Galbraith calls this organization the "techno-structure." This techno-structure serves several necessary functions. Because economies of scale require large pools of labor, individual worker freedom must be subordinated to corporate management objectives. Experts are placed at the head of each division to facilitate greater speed and efficiency in decision-making. This entire corporation, argues Galbraith, forces the individual to adopt, identify with, and become motivated by company directives. The company becomes a unified productive process. Promotion and recognition depend upon adapting to the prime directives. The same must be said for economic desires in the market. They must be manipulated in excess of needs so profits can

meet capital requirements.[7] In fact, this manipulation is really seduc-
tion because we believe we are "free to choose," to use a phrase
made famous by noted economist Milton Friedman. Glutted consumer markets mean unemployment. Mass production *must* mean
not only production and consumption by the masses but massive
consumption by consumers and producers.

Ellul argues that technology defines humanity: *Homo faber,* man
the tool maker; or Information Processing Systems, merely part of a
complex computing network. These mechanical metaphors remind us
of our degradation brought on by modern technique. Furthermore,
technology defines the universe we live in by believing that laws of
cause-and-effect govern reality. A god viewed in these mechanistic
terms loses the ability personally and providentially to sustain the
creation. Instead an impersonal deus ex machina, a god from the machine, governs. If we think, continues Ellul, that the main branch of
the Judeo-Christian does not understand God in these mechanistic
terms, we are mistaken. We understand evil as inexorable, just as we
understand technique as irresistible. Ellul goes so far as to argue that
technique is so totalitarian that it offers successful resistance to God's
counterveiling love.[8] "The dawn of the enslavement of the worker, the
destruction of the environment, and the bombardment of the consumer with a world of gadgets: these unmitigated evils are proof that
God has not chosen to resist evil."[9]

The modern state enhances the scope and the ideological impact of
technique. The state can use various techniques to control the military, administrative, and social sectors of culture. This is the heart of
the Republican critique of big government. "The basic effect of state
action on technique is to coordinate the whole complex. The state
possesses the power of unification, since it is the planning power par
excellence in society."[10]

The nuclear industry represents a case in point. The federal government funded the initial research for the splitting of the atom as
well as the "Atoms for Peace" project. They hoped to create a major
source of cheap, abundant power through harnessing the energy released by the splitting of the atom. The problems of clean-up, burial
of wastes, leakages, and explosion have demonstrated that the splitting of atoms has not been an entirely "peaceful" project, nor has it
been a local one. Indeed, the government has furthered techniques'
powers by strengthening the bureaucratic and administrative systems
that first funded splitting the atom and now must attempt to clean up
atomic wastes. Multiplying technique used to contain decade-long

leakages at Hanford Nuclear Reservation in the state of Washington is the best example of unpeaceful atoms: technique (bureaucracy and equipment) was created to put Hanford on-line, and then technique was used to clean up fission's mess.

Technique is, moreover, ecumenical; it canvasses the whole Earth, remolding different cultures into one more-or-less unified world system. Protestant/Catholic, capitalist/communist, Buddhist/Muslim— all societies fall prey to technique. Therefore, as technical objects are transferred from "First" to "Second," to "Third," and now "Fourth World" countries, ideology travels with it.[11] That is, technique transforms the social fabric to meet its needs. Nowhere is this fact more apparent than in the social transformation that caused the Industrial Revolution. Indeed, the very designation "First," "Second," and so on, suggests that the degree of technical sophistication contributes to a nation's international standing, self-identity, and place of importance.

Technology is autonomous; it is a law unto itself.[12] This much and more needs to be said. It is self-augmenting, self-replicating, and beyond human control.[13] Technology reinforces itself. Robots replace workers on assembly lines. Technique, argues Ellul, is self-directing. Technique directs reality. Everything must be made to conform to its dictates. The effect of the automobile is a good example. Traffic lights direct our pathways. Congestion crowds our highways. "Gridlock," or technique that chokes movement, is the result. There is a demonic irony to the word *auto-mobile* or "self-mobility." There is reduced freedom or mobility left on too many of our roadways because of the automobile.[14]

Since the Enlightenment, much of Western civilization has believed itself to be autonomous. We have defined ourselves as people who are self-directed. We invent, therefore, automobiles to enhance our self-mobility. Having become widespread, our "mobility" degenerates into death and gridlock. That is, we are forced to surrender our freedom, in auto deaths, congestion, and related losses, to our vehicles of mobility. Technique is monistic—that is, *one* universal force. It is one all-embracing power ecumenically uniting what appears to be divergent ideologies. Furthermore, technique is monistic in the sense that it acts as a force that blinds us to reality, and thus appears to be more clear-sighted than human Reason.[15]

Humanity has sought to secure freedom by the rationalization and mechanization of nature. We have robbed ourselves, ironically, of our freedom and authority, through the projecting of our alleged autonomy onto our machines, methods, and systems. While seeking to be free, we have become enslaved, argues Ellul. While wrestling nature for our freedom, we have subjugated ourselves to our technique. Perhaps the optimist might call for a technical solution to this loss of freedom. Perhaps a "fun-filled" vacation to Disneyland may help? The microchip, furthermore, has not brought a resurrection of democracy, claims Ellul. It has eroded our privacy as our government and our credit-card companies increasingly control and sell our private data.[16] The "information society"—saturated as it is with every form of information—nevertheless lacks wisdom. This society, dripping with information, foolishly must define people as information- or data-processing systems.[17]

TECHNOLOGICAL DETERMINATION?

One might conclude from this discussion that Ellul is a technological determinist. Is Ellul arguing that society and persons are conditioned, influenced, and defined totally by technique, as determinism indicates? He argues seemingly for technique's totalitarian effects. Technique seems to have left life barren. Is Ellul a determinist?

Ellul's theological works may not be omitted if one is to deal seriously with the question of determinism. The theological works, the second and equally important part of his thought, have different presuppositions. His theology contains thoughts on ethics and philosophy that match the weight and import of his sociology. Taken together, they form the foundation of his thought.

I am arguing, after the lead of D. J. Wenneman,[18] that Ellul's essential "method" is dialectical in character. It may be remembered that one finds in Marx's thought a dialectical method of interpreting history. Accordingly, two contradictory forces—the revolutionary class-created thesis/antithesis—form the dynamic of history. Ellul also makes use of a dialectical method of interpreting history, though his is not of an economic nature. Ellul's essential dialectic revolves around "two totalities"—that is, two mutually exclusive, all-encompassing ways of thinking form the codeterminants of Ellul's thought.[19] "From the beginning my thinking revolved chiefly around the contradiction between the evolution of the modern world and the Biblical content of revelation."[20] This "evolution" Ellul speaks of is best portrayed by the thought of Karl Marx. Thus, methodological Marxism and Christianity are the two totality ethics that together and in contradiction form the basis of Ellul's thought.

I realize that Christianity was a totality implying an ethic in all areas, and that Marxism too claimed to be a totality. I was sometimes torn between two extremes, and sometimes reconciled; but I absolutely refuse to abandon either one. I lived my entire intellectual life in this manner. It was thus that I was progressively lead to develop a mode of dialectical thinking which I constantly made my foundation.[21]

The contradiction between methodological Marxism and Christianity may be expressed as follows. Marxism is materialistic, prone to determinism, and holds autonomy as an important human virtue. Christianity, on the other hand, incorporates the goodness of the material world into a more holistic picture, is more decentralized, less myopic, and is heteronomous in nature.

Ellul makes no attempt to synthesize these two contradictory yet "totality" ethics. Indeed, to do so—say, in the form of a "Christian" worldview—would be to create an ideology or another totalitarian method of control. Contradiction is a must. Contradiction produces tension. Tension is necessary because it produces action, an action that addresses technique's problem. We have the possibility to address this tension because— and this is key—we are capable of responsible free decision. It is an existential, authentic human encounter with technical reality that necessitates a free decision.

One may ask, what does all of this philosophy have to do with technique? It is by developing this dialectic that Ellul is able to carry out his task of relocating humankind within the modern technological universe. What is lacking in such a universe is precisely that openness that allows for the historical development of the free subject. For technique, in Ellul's view, represents a closed world. Its development is completely mechanical. Its results are completely predictable. Ellul's dialectic itself introduces a certain negatively and thus unpredictability into this technical system. It represents a negation of technique.[22]

The last phrase, "negation of technique," signals a new facet to Ellul's dialectic. Ellul's system of thought fundamentally consists of technical necessity on one hand and the possibilities of human freedom on the other hand. Freedom represents an absolutely essential condition for human existence.[23] The human spirit longing to be free arises like an unstoppable force against the immovable object of technical necessity.

Ellul believes the truth of Christian revelation must confront the sociological world dominated by technique. The truth that confronts technique is an alien truth to this world. While alien, this truth, nevertheless, *is capable in some unfathomable way to contradict the way of servitude*. This way of contradiction, however, is rarely (if ever) realized in human history.

Thus, there exists a profound dialectic in Ellul's thought: the existential, theological need to be free is pitted against the sociological, technological necessity of a determined society. A second dialectic also exists. The sociological side of Ellul views the world through the glasses of empirical, evolutionary reality. Accordingly, the visible, verifiable, measurable way of seeing reality predominates. On the other hand, Ellul views reality as a theologian. Faith, the grip of divine revelation, and supramaterial reality are the dominant presuppositions to this way of thinking. Taken together, though never synthesized, the principles of the empirical and the principles of transcendent reality form the second dialectic in Ellul's thought. Faith versus empirical seeing, evolution versus unfolding, verifiability versus faith in the unseen, and sociological facts versus theological beliefs form the second dialectic in Ellul's thought. Again, no synthesis is allowed.

ELLUL'S THEOLOGY

Talk of transcendent reality naturally leads me to a discussion of Ellul's theology, the core notion of which is freedom. In a moment I will mine Ellul's notion of freedom as an antidote to technological necessity. For now, I must probe more deeply into the origins of Ellul's theological beliefs. Are there the forces capable of lifting Ellul and the rest of humanity from the depths of technological determinism?

Ellul believes there is no activity around which all of life should revolve; there is no absolute in this life. Everything and everyone is dependent or contingent upon God who transcends all given reality. Therefore, it is not possible for humans to absolutize or make a universal truth out of anything or any process.[24] Thus, Ellul says in "faith" that technique cannot and should not be absolutized. Faith addresses God with the full confidence that evil is not sovereign, is not absolute. Belief, on the other hand, represents confidence or trust in something or someone other than God. We believe the world is not flat because we have been told it by those whom we trust. Belief is concerned, moreover, with "religion." "Religion binds people together and binds them as a group to their god."[25] Reason, science, technique, or money can be deified. Belief is thus false faith, according to Ellul.

Wanting to resist false faith, Ellul defies technique with a prophet's wisdom. Thus, Ellul argues that progress (as understood in chapter 1) is presumed by modern secular society to be a virtue and thus gains adherents.[26] This secular yet religious notion of progress is supported by the belief in human reason and its power to dominate nature in the name of freedom. While claiming to set us free, this false religion does not make good on its promises. It leads to bondage and to oppression, a loss of personhood. Belief always tries to construct something or trust in some force it believes to be a god.

Ellul places himself, naturally, at odds with the technological optimists. They seek their earthly salvation or fulfillment in technique; Ellul sees demonic servitude. The optimists praise humanity for technical progress; Ellul sounds the prophet's lament because humanity has become captive to technique. They celebrate reason, autonomy, and progress; Ellul mourns because of this counterfeit trinity. The optimists believe that humanity will eradicate life's perplexing problems through the use of technique; Ellul believes the destruction brought on by technique greatly exceeds its goodness. In short, Ellul is a pessimist when he contemplates that technique has and will contribute to the human condition.

Ellul argues for the meaningfulness of a life defined by false faith, as a Christian would. Technique has not made life absurd. Mindless consumption, the nuclear arms race, or suburban shopping malls—all particular manifestations of technique—are absurd. These particular absurdities do not constitute a meaningful life. Life is not oriented to nor does it find its significance in technique, is his Christian confession. Life is given meaning by God that technique cannot altogether eradicate.

On what basis does Ellul establish his view that life has meaning? The answer to this question is crucial as it represents the core of Ellul's theology.

> I myself have been gripped by the unique and irreplaceable character of the word, but for very different reasons: because God created the world, because he has revealed himself uniquely by His word, because the incarnate Word is the Word of the eternal God, because the God in whom I believe is Word . . . there is order and truth in reality.[27]

Though technique does overpower us on the human level, God's word is not subject to this domination. God's word is free to address us because it is a transcendent word. This word "above all earthly powers" (to use Martin Luther's famous phrase) speaks to us and for us and thus provides the basis of freedom from the demonic powers of technique. We find the clarity of mind and will to see reality as it is when we are addressed by the Word made Flesh, Jesus Christ. The consequence of this vision is the power to act freely and responsibly. Thus, "enlightened" by God's word, Ellul can understand the difference between, for example, the relative verbiage of computers and the clarity brought by God's word. Thus, "computers can understand human phrases relating to acts and limited objectifiable concepts. They can give information and obey orders. But they plainly have nothing whatever to do with word or speech."[28]

Ellul's understanding of dialectical thought originates in his view of God's relationship to the world. Dialectical contradictions often are manifest because God's word as one basic reality is often contested by evil—in this case technique. God's word can summon dialogue, or it can engender contradiction or cause distance between God and the one addressed. The word of God—our grace—contradicts our sin—or technique. Ellul wants the reader to engage in a dialogue with God's word because of the sin of technique. We are called to dialogue, to regain true communication, in the midst of the computer age because of the idolatry of technique. Our sinful nature inclines us to idolatry, which is the false but seductive attempt to make absolute that which is finite in the creation. Because of the interplay of sin and God's word, reality is inclined to manifest a two-sidedness, a yes and a no, a positive and a negative, grace and sin. The positive and the negative, sin and salvation, do coexist; they do not rule each other out. Ellul confesses that each force is essential for the other's continued existence. Thus, the evil of technique is necessary for the good of redemption. "Negativity is essential, for if the positive remains alone, it is unchanged, stable and inert. A positive element, for example, an unchallenged society, a force without counter force . . . is enclosed within the permanent repetition of its own image."[29] Ellul sees this dialectical as inherent to reality.

The root of Ellul's pessimism can be located in his dialectics, as this key quote says:

> I might say that it is a dialectical attitude that leads us to consider that we are impotent in relation to structures and necessities but that we ought to attempt what can be attempted. The same attitude causes us to affirm constantly that as an expression of determinism and as an exclusion of freedom, society must be unceasingly attacked and yet that all our efforts will tend to maintain this society. . . .[30]

The "structures and necessities" refer generally to sin and specifically to technique. This statement taken alone may lead one to conclude that Ellul is a technological determinist. However, it must be interpreted within the rest of Ellul's writings. Ellul speaks in the next chapter of this same book of "harmony," of a "correspondence," a joy, a simultaneity of occurrences, a fullness of being that surely contradicts the bleak, myopic picture painted in *The Technological Society*. Thus, salvation—harmony, correspondence, joy, simultaneity, a fullness—for life's many activities seems to be available, though the prophet never tells us the road to repentance.

The reader must not be confused at this point. After having set before us the overwhelming evil of technique, Ellul confronts us with the all-encompassing possibility of redemptive harmony. Taken together, these two contradictory ideas form a dialectic, a dialogue between contradictory statements, the purpose of which is to move us to action, to resistance of evil, to confrontation with technique.

Nevertheless, and this fact is crucial in understanding Ellul's dialectic, while technique actually rules history, harmony represents our hope for (not in) history. Ellul's lack of specific examples of harmony leads to confusion and uncertainty as to what he is proposing. "Let us return to the earth and try to make it humane, livable, and harmonious. This is our real business. . . . Let us rediscover the earth in joy."[31] We are asked to return to Earth—the place of the domination of technique—but find there no literal place of freedom or a program of action.

I must further clarify Ellul's notion of freedom. One may thus far conclude that Ellul minimizes the human ability to resist evil. Consequently, we can experience little or no freedom because technique dominates history. The question continues: is Ellul a historical determinist or is he not?

A complete understanding of Ellul's dialectics demand that as much attention be given to the notion of freedom as has been given to technical necessity. This is a difficult task because Ellul has said technique is a universal, monistic, ecumenical or global, and self-augmenting force. Therefore, the freedom he is about to define must be equally fundamental to human existence, as it must bear the enormous burden of providing a counterweight to technical necessity. "Freedom is the ethical aspect of hope. An ethics of freedom can be found only in hope and can only try to express hope."[32] Freedom does not spring out of hope by a kind of necessity. Freedom is the real and authentic possibility of choosing one's destiny. Freedom is created by God for humanity in The Human. The Human is Christ and His hope is the salvation offered by Jesus Christ. Hope for and in history comes after the response of man to God's love and grace, while freedom is the gift of God to man's hope. Freedom and hope are absolutely essential characteristics of humanness. In Christ we realize them; apart from Christ we long for them. Ellul believes that God is the Author of freedom. Freedom is *the* essential characteristic of both humanity and God.

God's existence is said to transcend that of our own. This fact is crucial when we remember Ellul's historical determinism. That is, God's glory, honor, and might exceed in might but are not unrelated to our reality. Hence, God is free from technique's enslavement. Ellul finds in the transcendent God the *only* source of freedom sufficient to rescue us from the gulag of technique. Since history is determined by technique, only a God who stands above and beyond history, who is not tainted by technique, can give us the hope of freedom. Only God can give humanity the possibility of a living hope in a world under the rule of technique, claims Ellul.

While God's existence is transcendent to that of our own, His[33] relationship to us is not. God is fully revealed in Jesus Christ. Christ, though fully God, is fully human as well. Christ enters into our world as a fully immanent link between humanity and God. It is the immanent link that forms the basis for the possibility of freedom from the vice-grip of technique. "Destiny has been lifted by the act of Jesus Christ." After him, "there is no more ineluctable necessity."[34] This "act of Jesus Christ" that Ellul is referring to is the destruction of all oppressive powers by the kingdom or the rule of God.

Freedom means the liberation from technique's domination. Because Jesus Christ is free from sin and death and because he is in us by his Spirit, we have the hope of freedom. This freedom releases us from alienation and reconciles us to all of life's vital and holistic forces. We have the *hope* of freedom, but do we have the reality of freedom? Freedom means that one is no longer possessed by anything external to oneself. Money, sex, fame, security, and, above all, technique may no longer rob us of our freedom. We are no longer *"Homo faber"* (as Marx said) or an "Information Processing System" (as modern computer experts say). We are free human beings: vital, alive, complex in being, and reconciled to oneself and to the world around us by Christ.

Freedom and hope offered in Christ exist not only for our personal lives, however important that may be. Rather, freedom is extended to

the entire world. Freedom rests on a hope that there exists the genuine possibility of an authentic future, instead of one determined by
the dialectics of history.[35] Thus, Ellul does not envision the dialectical
way of being as eternal. Dialecticalism is the necessary accommodation to the evil of this world—preeminently technique. We are being
led, he continues as a theologian, to a world that will become a heavenly Jerusalem,[36] a world freed from evil and dialecticalism.

Ellul's emphasis is not oriented totally to another world. He sees
that this world is also important. However, and this note is crucial,
when we press him for concrete details as to how his view of freedom
can positively affect history, Ellul says there can be no Christian factory or a Christian philosophy; no concrete alternative. That is, no
historical event or project that may be labeled Christian, because to
do so would be a supreme act of arrogance. Whenever Christians
have tried to identify some project such as the Crusades as Christian,
it has only served to enhance an ideology of oppression, claims Ellul.

Ellul, the social critic cannot speak of freedom for too long. He
must return to the theme of "determinations." There are social forces
at work that oppose our need for freedom. Economic alienation or
class hostilities represent the first determinant. Second on Ellul's list
are sociological determents. Urban environment, organizational techniques, mass media's manipulations, and the expansion of the state
are among the social determinants. Indeed, the modern state has
eroded the freedoms of democracy by its manipulation of voter pref-
erences. However, just as the reader again begins to think that life is
socially determined, Ellul dialectically reasserts the need for freedom.[37]

What meaning can freedom possibly have for the historical world
of necessity or determinations? "Freedom has meaning only in relation to an authentic necessity. Freedom is fate overcome, an obstacle
surmounted, a limit passed, a sacred sphere secularized. . . . Freedom
loosens up tightly regulated mechanisms. . . ."[38]

Freedom/necessity: the **Scylla and Charybdis** of modern existence? I am haunted by the question at this juncture of Ellul's
thought as to how relevant freedom is for real history. Are the terms
hope, freedom, and heavenly Jerusalem historically meaningful to
any degree?

Perhaps Ellul's more recent works will answer this question. Has he
softened his critique of technique? Do we live in a society that is technically determined? The aged lion is not toothless. In *The Technological Bluff*, Ellul writes, "My warning today is the same as 1954, when

I wanted to alert people to the future potential of technique and to the risks entailed by its growth so that they might be able to react and to master it, lest otherwise it escapes their control."[39]

As in *The Technological Society*, the master critic of technique here wants to expose the deception, the overconfidence, and the misleading nature of modern technique. The modern technological bluff consists in rearranging everything to accomplish technical progress. Politicians manipulate media images to create the illusion that your cause is their cause. The media shower us with pseudo-images trying to convince us that this, that, or the other product will fulfill our deepest needs. The economy is manipulated like gears in an automobile so that it can produce the kind of economic growth we desire. All of these particular methods of manipulation (and many more besides) have heightened the mesmerizing narcotic of modern technique.

Ellul continues: Prior to 1950 humans lived by the imperialistic metaphor of the Industrial Revolution. Starting in the middle of the eighteenth century, man, machine, nature, and nurture[40] were all coordinated into an efficient productive system of wealth production.

Today the technical metaphor has changed but the oppression has not. The information revolution has produced the new imperial metaphor of the information society. Information is our new environment. We awaken in the mornings to our clock-radios. We listen for the weather and traffic conditions (the latter often oppressive, especially in larger cities) on our car radios. We then move to work where a computer, a fax machine, and mail inundate us with more information. We return home to our television sets sometimes to find a few dozen or so channels flooding us with information. Finally, we may close our day by reading books. Ellul, therefore, argues that we live in an information-soaked society, a society that, nevertheless, is lacking in wisdom and ethical sensitivity. Thus:

> I refer to the fact that technique is our environment, the new 'nature' in which we live, the dominant factor, the system. I need not elaborate on its features: autonomy, unity, universality, totality, automation, causal progression, and the absence of finality.[41]

So the prophet sounds the same warning today that he did more than thirty years ago.

Ellul carries the battle to the heart of technological optimism. He critiques the foundations of Julian Simon's notion of linear progress, calling Simon's thesis "absurd." Simon believes the economic market to be a place of pure and perfect competition. Supply and demand should automatically balance and will consequently produce optimal social conditions; so the myth goes. Workers, technical innovations, and capital sources will combine to eliminate all long-run (hence linear) concerns. This again is Simon's argument.

Ellul counters: Even if optimal resources were to be allocated—and Ellul doubts they will—the resulting market structure that regulates this "brave new world" would be totalitarian in nature. People caught in this omnivorous economic organism would become one un-differentiated resource passively waiting market manipulation and exploitation.[42] Indeed, this has been the legacy Western capitalistic imperialism has left the world.

Second, Ellul argues that Simon's facile optimism knows no social limits or ethical norms beyond economic growth. "The optimism of this economist rests, then, on an absolute belief in unlimited progress. Whenever a difficulty arises, 'technical progress will deal with it.' We have here an absolute form of the technological bluff."[43]

Economic progress, under these conditions of optimal economic growth, becomes social regress or social retardation, according to Ellul. The different aspects of society, such as education, government, and the natural environment, lose their integrity to the god of tech-nique, so the prophet argues. With that note, Ellul rests his case—the case for our defense.

Pessimism Continued: Neo-Marxism

One must not gather the impression that Ellul is the only philoso-pher who is pessimistic about technology's impact. The twentieth cen-tury spawns pessimists because of its technological nature. For example, the Marxist school of thought shows additional strains of pessimism.[44] It was noted in the last chapter that Marx was optimistic about the future of technology. This optimism is not shared by one of the chief modern disciples of Marx.

Jürgen Habermas stands as one of the luminaries in neo-Marxist thought. His work *Technik und Wissenschaft als "Ideologie"* (*Tech-nology and Science as "Ideology"*) is his major work on the subject of technology's effect on modern society. Unlike Marx, Habermas does not envision the workers taking control of technology, removing the alienation that exists between themselves and the works of their hands, and realizing a bounty of goodness. Rather, "the liberating power of technology—the instrumentalization of things—is perverted into a fetter of liberation: it becomes the instrumentalization of man."[45] Habermas argues that modern technology has chained us to instrumental reason.

I showed in the first chapter that optimists exalted instrumental rationality because it was thought to conquer nature and, consequently, ensure economic rewards called utilities. Habermas, however, laments that humans have become *things* or coordinated parts of the production process. This is certainly true for the worker Marx wanted to liberate. Workers have lost their freedom and self-determination to technique. The modern assembly line testifies to this fact. Since the Industrial Revolution, humans have been forced to adjust their efforts to the dictates of the assembly line when they were employed. Today, robotics—the branch of technology that has mechanized and routinized human labor—has taken the jobs of many assembly-line workers.[46] Nowhere is this more apparent than in Detroit, the former capital of assembly-line labor.

It is noteworthy that a neo-Marxist like Habermas would use the word **ideology** to refer to technology. Marx argued that the capitalistic ideology dominated society and worker control of technique was the only hope for salvation. Technology, now used as a means of control by the capitalist, has become the modern origin of human oppression, according to Habermas.

Habermas views the modern state with more cynicism than did his ideological father. The state increasingly intervenes in society to assure economic progress by co-opting science and technology and then by using them for economic ends. The result is that science and technology have become the means of state domination and hence oppression. Nowhere is this more apparent than in the billions spent for weapons procurement. The optimistic Marx believed a state-owned technology would bring Utopia; the pessimistic Habermas believes the modern state-controlled technology only serves to oppress us. Habermas wants nothing to do with state domination.

Modern technology and science dominate or oppress people in the name of "progress." This "unfreedom" appears neither irrational nor totalitarian but appears in the guise of leisure, wealth, and increased status.[47] This statement is more true now than when Habermas wrote it. Leading journals in science and technology are calling for increased federal involvement in science and technology—and the U.S. government is listening. In fact, some are noting the emergence of "a U.S. 'technology policy,' in which the federal government helps develop and provide access to the technical knowledge that will make the American economy more competitive."[48]

As politics surrenders its integrity to science and technology, the democratic will to resist technology's domination lessens. People are hypnotized and drugged by technology's effect. This narcotic effect is certainly true of our existence in front of the "tube." Televised political debates give the appearance that the electorate are viewing debates involving substantive issues vital to our democratic future. In fact, we are witnessing the manipulation of images so that the electorate feel satisfied that their candidate thinks as they do. This manipulation and unreality is apparent especially at national political conventions held before elections. Substance has vanished. The legacy of actor Ronald Reagan's image-conscious presidency is ample testimony to this fact. We are "hooked" on images.

Habermas argues that we do not enjoy autonomy; we must settle for quasi-autonomy.[49] We believe we are free to consume products and enjoy leisure as we please. In fact, "industrially advanced societies seem to approach the model of behavior control . . . by reconstructing [our lives and nature] after the model of self-regulating systems of goal-oriented and adaptive behavior."[50] Accordingly, "technocrats," or technologically advanced experts, are consulted for advice that can be used to manipulate behavior. Advertising executives know that technically choreographed sex sells blue jeans. So they place two young, well-endowed members of the opposite sex in a "bump and grind" position and the message is clear: wear these jeans, engage in these actions, and your sexual fantasies will come true. Thus, the ideology of technology determines our lives and corrupts our freedom. Marx failed to realize our enslavement to technique, claims Habermas.[51]

Habermas opposes the domination of what we have called instrumental rationality or reason used as a means to the end of manipulation and control. He believes that communication and social mutuality must replace the reign of impersonal, manipulative instrumental reason. Only then will we become "dominion-free," or freely emancipated individuals. Accordingly, communicating and interacting would not have control as its goal. Rather, individualization and choice would become the new virtues.

Habermas, like Marx, wants a revolution. Unlike Marx, Habermas's revolution does not involve alienated workers. Rather, circa 1969, Habermas appeals to students who have prospered because of economic and technological progress. These students from high-income, highly motivated families and who are not themselves the product of authoritarian homes must be enlisted for this "revolution

of consciousness." These students are the locus of his hope and salvation. These are the new social elite who, because of their privilege, can distance themselves from their peers and thus form the
basis for protest. These elite will protest the technocratic "achievement ideology" of Western life. These students can turn to a thorough critique and renewal of technology because their wealth,
status, and eventual success are assured. This assurance gives them
the aplomb to guarantee a new future free of technical domination.
This grand vision is predicated upon the wealth and leisure that
technique brings.[52] Technological fruits make a new order possible,
claims Habermas.

The last point needs to be emphasized. Technological fruits give
elite students the resources necessary to take distance from technique
and thus form the basis of critique. Habermas does not seem to be a
dire pessimist like Ellul. Habermas's pessimism seems to be moder-
ated by his need for the fruits of technology and for his faith in the
young. Without wealth, leisure, and status, student protests may
never occur.

I label Habermas, therefore, a secular conflicted pessimist, though
without Ellul's dialectical methodology. His conflict centers around
his repudiation of instrumental rationality, the heart of technical activity. On the other hand, Habermas needs the fruits of technology to
give students the resources necessary to critique technology. His views
are in conflict but are not dialectical because no methodological inner
point of thought integrally links inherently antagonistic forces as it
does in Ellul's dialectical methodology.[53]

Pessimism Continued: A Representative Christian Theologian

A more subtle form of pessimism awaits us. The work of Nicholas
Berdyaev in *The Bourgeois Mind and Other Essays* is an example of
soft pessimism. This Christian thinker can speak of the powers of
technique as "absolute," "the thing placed above man," and capable
of bringing destruction to culture.[54] He seems to give it an autonomous character when he says that "technique knows no symbols;
it is realistic, reflects nothing, creates only new actualities . . . ; it di-
vorces man from nature and from others."[55]

God, on the one hand, creates organisms that are interconnected
with, and the foundation for, other forms of life. On the other hand,
organizations such as bureaucracies are artificial because they are

constructed and, hence, cannot spawn life. The difference between organisms and artificial constructs is that living organisms both create and are spawned by life, while constructs are lifeless. "The supremacy of technique and the machine is primarily a transition from organic to organized life, from growth to construction."[56]

Life that is centered only on constructs, as modern civilization is, produces "hopelessness" because a "new cosmos of its own creation" with unforeseen consequences attempts to pass as life. Modern, fabricated, constructed life mistakenly passes as vibrant life. The origin of this alarming condition is the alleged autonomy of the machine. This machine spells the ultimate end of man and nature. Thus,

> the wireless will be carrying the sound of music and singing and the speech of the men that once lived; [nevertheless] nature will be conquered by technique and this new actuality will be a part of cosmic life. [Therefore], man himself will be no more, organic life will be no more—a terrible utopia![57]

Nevertheless, Berdyaev is not a strident pessimist as is Ellul. While noting technology's potential or real evil, he does not absolutize human freedom or evil. He ties the machine to human responsibility, which in turn results in at least two consequences. If machines are not laws unto themselves, then they may be made to be subject to other laws and customs that could regulate their existence. If machines could be made to obey external laws, then they can be made to be responsive to human demands. They can serve in a way that frees rather than enslaves us.

> It is not machinery, which is merely man's creation and consequently irresponsible, that is to be blamed . . . it is unworthy to transfer responsibility from man to a machine. Man alone is to blame for the awful power that threatens him; it is not the machine which has despiritualized him—he did it himself.[58]

Berdyaev halts his slide into strident pessimism by linking the machine to human responsibility. Machines are not, therefore, autonomous but are tied to human responsibility. Humankind has the *potential* to control the machine.

Nevertheless, it is appropriate to speak of Berdyaev's pessimism because he uses such words and concepts as "absolute," "destruction of the culture," and "divorce from nature" when referring to technology's effect on contemporary life.

POSTMODERNISM AND TECHNOLOGICAL PESSIMISM

Perhaps the most relevant way to bring this analysis of technological pessimism to a close is to examine the critique made by postmodernism. The term *postmodernism* refers to a loose collection of thinkers who take intellectual distance from "modernity." Modernity means the intellectual, social, and religious project initiated during the Enlightenment. The postmodernist's critique centers on modernity's attempt to subjugate nature by Reason in the name of human freedom.[59]

The postmodernists critique the methodology and the product of modernity. Accordingly, ideologically neutral knowledge produced by detached scientists is passed along to engineers whose products are believed to be inevitably beneficial. Therefore, technical expertise is thought to be the conduit through which human life is bettered. This betterment, as I have shown, is called progress. Thus, "a faith in the powers of knowledge and technology to ameliorate human life and solve the basic problems of modern society has been one of the central features of Western culture."[60]

Mounting environmental problems, changing philosophical paradigms, and evolving views of the nature of the person have caused increasing doubt to be cast on the Enlightenment's view of the person as a rational, utility-maximizing lord of nature.[61] These fundamental changes have led to a search for a new kind of technology built on new principles and guidelines. Accordingly, many schools within postmodernism stress a oneness with nature, personal wholeness, pastoralism, and rejection of the crass economic materialism of capitalism and Marxism.[62] Thus, noted postmodern thinker Leo Marx says,

> Those hopes of modernity were grounded in what postmodern skeptics call foundationalism: a faith in the human capacity to gain access to a permanent, timeless foundation for objective, context-free, certain knowledge. The stunning advances of Western science and the practical arts seemingly confirmed the epistemological faith, and with it the corresponding belief that henceforth the course of history necessarily would lead to enhanced human well-being.[63]

Perhaps Lewis Mumford was the first postmodern popular critic of modernity.[64] His architectural criticisms were leveled against "the belief in mechanical progress" that made technology an end rather than simply a means. Mechanical progress, argues Mumford, determined all aspects of the building practice. When Mumford first made this criticism in 1962, he was part of a distinct minority of "marginals" who did not fully understand the alleged benefits of modern technology. Mumford's objections gathered increasing support in spite of marginalization. His critique was aimed at the destructive effect of technology on moral and political goals.

America's involvement in Vietnam caused the criticism of technology to accelerate. It was an era of immense disillusionment with technology, economics, and much of American society. Criticism was aimed at entrenched power centers for the way they had corrupted American democracy. Centers like the Pentagon[65] used electronic communication systems to gather information on friend and foe alike. Critics wanted a more democratic, less centralized form of technology. Optimists argued that more democracy was on its way with the advent of the modern computer. However, it was the computer, the postmodernists countered, that was enhancing the concentration of power in the sixties. Revelations about how the Nixon White House used the computer to store data on innocent citizens proved the pessimists correct.

By 1969 pessimism became an established and respectable criticism. In March 1969, at the highly touted Massachusetts Institute of Technology, the Union of Concerned Scientists convened a meeting to challenge the current position of science and technology. These noted scientists criticized America for misusing science and technology to such a degree that the destructive duo represents a threat to the modern existence. The breadth and the force of the resulting communiqué was historic. While focusing on the technology of the Vietnam War, the scientists went on to make a more blanket condemnation of modern technology and science citing its threat, waste, and totalitarian nature.[66] This statement marked the beginning of the "counter-culture," because the dominant culture—the science and technological establishment—as well as its values—progress, materialism, and objectivity—became the object of searing criticism.

Postmodern criticism erupted in the hallowed halls of science. In 1962 Thomas S. Kuhn published *Structure of Scientific Revolutions*. He argued that "truth" inevitably is viewed as an evolving, consensual product that becomes socially accepted through what he called a paradigm. This consensual mass of scientific knowledge was in structure and content the exact opposite of the Enlightenment's view that truth was objective, value-free, individualistic bits. Paradigms were subjective, consensual, and collective.

Kuhn's work caused an explosion of intellectual controversy. Later in the 1960s another author dropped another intellectual bomb that shocked more than the technical world. In 1964 Herbert Marcuse published *One Dimensional Man,* in which he criticized the "technological universe." Technological rationality, he argued, forms a comprehensive and powerful mode of discourse and action that has dominated and subjugated not only nature but humanity as well. Social control brought on by "predatory affluence" has all but finally eroded the essential freedom of mankind.[67] Modern technical experts clearly are repudiated in his seminal work *Essay on Liberation.* Thus:

> It is still necessary to state that not technology, not technique, not the machine are the engines of repression, but the presence in them of the masters who determine their number, their lifespan, their power, their place in life, and the need for them. It is still necessary to repeat that science and technology are the great vehicles of liberation and that it is only their use and restriction in this repressive society which makes them into vehicles of domination.[68]

Note must be made of Marcuse's ambivalence. One the one hand he argues that we may not blame technology, but on the other hand they may become "vehicles of dominations." At the same time he notes the servitude brought about by modern technology. Thus, a call for liberation had to go out.

Theodore Roszak added fuel to the social fires of criticism that burned in the 1960s when he published *The Making of a Counter Culture* in 1969. Roszak takes aim at the myth of "objective knowledge and consciousness." This type of knowledge was believed to be value-free and without ideological taint. Science and technology has become the dominant ideology because it has absolutely saturated our society, argues Roszak.

Furthermore, he vigorously attacks the technocracy. The technocracy is the regime of experts whose scientific and educational credentials and expertise allow them to assume an unquestioned role in the deployment of technology. Trust in these experts leads to an erosion of our essential human freedom. This technocracy is antidemocratic, coercive, and self-legitimating because of the degree of control we place in their hands. We surrender our freedom to them because of the number of comforts and securities afforded by the experts. Consequently we are dependent on technology for our vital needs. Industrial affluence is the narcotic that soothes the hardly recognized but profound pain produced by the loss of freedom.

Roszak thought he had found in "anti-intellectuals," or countercultural youths who practiced shamanistic rituals and unhindered sexuality, who lived in utopian and romantic natural communes, the élan of liberating salvation for a technically enslaved culture. Sadly the counterculturals of the sixties became the "me generation" who were slipping and sliding down the fast-track of economic satisfaction during the 1970s and 1980s.

Postmodernism's alternative world view, which may have lasting value, seems in many ways diametrically opposed to modernity's vision. An ecological consciousness that stresses the links and associations, rather than the dissimilarities and domination, between humanity and nature is key. An organic rather than a mechanical view of the cosmos thus emerges whereby considerable natural autonomy seems the dominant principle. Accordingly, self-regulation, self-correction, and self-propulsion are the autonomous principles touted by the postmodernists. Thus, "if mankind is to escape its programmed self-destruction, the God who saves us will not descent [*sic*] from the machine: he will rise up again in the human soul."[69]

It needs to be noted, in conclusion, that both postmodernism and modernity share the same foundational principle: human autonomy. Note should be made of this fact for it will play a role in our last section on discernment.

THE PLACE OF TECHNIQUE

It follows from what I have shown that pessimists would have little to do with modern technology as it represents one if not the principle problems for modern life. At the extreme, Ellul would abandon it altogether if he could. He would return to a small group experience of interpersonal communication, thereby seeking to revive freedom and meaningfulness for life. This is the best alternative that his dialecticism has to offer.

Less dire are the postmodernists. However, their ambivalence leaves them with no viable alternative to the technology they so brilliantly critique. Roszak's shamanistic young who then were thought to signal the dawn of a new age now seem, in retrospect, self-destructive and self-indulgent.

These faults underline my point. Worldviews, principles, and ultimate commitments do shape the way one looks at life. A shamanistic, autonomous, secular view was the one chosen by Roszak to represent his faith in a new technologically free future. Technology was best spurned for higher levels of consciousness reached through drugs and magic. At the same time, viewing life does mean locating technology within some grid of experience. It means locating, placing, or orienting technology within some meaningful context of life. To say that Roszak wanted nothing to do with technology is to say that viewing is taking place. To say that *no place* or only a *small* place should be allotted to technology is locating technology *outside* or, minimally, *alongside* the boundaries of human experience. Modern technology must go away or shrink, cry pessimists, because technology itself is the source of the woe of modern life. However, *no* place literally cannot be found because it does not exist. Even if we are generous, beyond what postmodernists say, and grant that they do not want to abandon technology but only relativize it, no plan for locating technology is in evidence. Hence, while postmodernists' criticism does advance the discussion, their silence about the place and the proper development of technology seems to doom them to become historic footnotes rather than cultural shapers.

Discerning Pessimism

The reader may be inclined to be cautious in accepting Ellul's or others' diagnoses of our technological situation. Given an inclination to moderation and aversion to extreme negativism, one may accept only some of the pessimist's thought. The individual experience of technology may not be so negative. People drive their cars and experience the freedom of mobility. They may have taken their children to Disneyland and not thought of it as the "ultimate idiocy." In everything from electricity to modern medical technology, our lives have benefited from technique—or at least they are not as doleful as the pessimists would have us believe.

This moderate position is, in my opinion, naive and simplistic. The Judeo-Christian tradition has always confessed that humanity was made in the image of God.[70] The image must never be cheapened by worshiping and serving—glorifying and modeling—any one aspect of or thing in the creation. When Frederick Taylor created a technique for workers by which the motions of the efficient machine were mimicked so that more economic rewards could be gained, our image was corrupted. That is, we have surrendered a measure of our dignity, freedom, worth, creativity, and responsibility to the dictates of the machine. Thus the pessimists are correct when they say, "the machine demands that man assume its image; but man, created in the image and likeness of God, cannot become such an image, for to do so would be equivalent to his extermination."[71] Machines are to serve humans, not humans the machine. When persons do surrender parts of their humanity to the machine, we must be alarmed.

The fact that people have not sounded general alarms may signal their duplicity. Perhaps we make such weak, shallow, abstract critiques of the machine because we are willing to exchange its alleged benefits for our freedom. Our lifestyles betray our real commitments. Pollution, gridlock, even deaths, that result from our steel chariots of "self-mobility" speak more loudly to a commitment to technique than do any glib clichés of moderation. As consumers, people are demanding lightweight, speedy, pollutant-emitting machines that provide the means to destroy life—quickly or gradually—at an unconscionable rate. It is blasphemous and irrational for us to mimic the machine because, however much we have in common with nature and with culture, significant parts of our image and abilities are unique. It is antirational to trust the technocracy because it threatens our freedom for responsible action.

Our dependence on technocrats, or experts trained in the efficiency of a given technique, betrays a weakening of our responsibility and community. The years of parenting expertise gathered by friends and neighbors can be as beneficial as an expert's advice for the parenting that my wife and I do. Should a consultant's advice be needed, seasoned parents should compare that advice with that of trusted friends. We can and must trust our judgment as much as we trust the advice of experts. We surrender a portion of our responsibility to technocrats when we refuse to trust our own judgement. To the extent that we have surrendered our responsibility to technocrats, and to the extent that pessimists resist this loss of self-determination, the pessimists advance the discussion through their critique of **expertism**.

Technique is universal. Its effects circle and emasculate the globe. These destructive effects were witnessed firsthand during my visit to the Third World country of Zambia. Between British colonial imperialism and capitalistic modernization, "primitive" Zambian traditions—beautiful and attractive in their own right—are fading. Our techniques have a mesmerizing effect that can be seen, especially on state-controlled Zambian television. The first program I watched was an hour-long documentary, put together by Zambians, about the evils of capitalism and socialist-Marxism and the need for a uniquely

Zambian economic ethic. Impressed by this statement, I wanted to sample more. The next program aired was a rerun of the American series *Miami Vice*![72] Technique is universal.

Technique has become a false ideology and therefore a detriment to our lives. To view God, as the mechanist does, as remote watchmaker unconcerned about the intimate affairs of Her creation is to miss the personal, providential, parent-like care of God, a confession Christians and Jews have made from our beginnings. To think the world subject to iron-clad laws of cause and effect, as Descartes[73] did, is to rob us of our responsible freedom. Freedom cannot exist in a deterministic cause-and-effect universe. Whenever the place and importance of technique is exaggerated in our lives, **idolatry** waxes and freedom wanes.

While arguing that a portion of the pessimists' critique is accurate, and while wanting to avoid a shallow moderation, must we say that the pessimist's discernment is final? I think not. Have you ever seen a machine build itself? Do machines repair themselves? When has a machine started itself? Do machines object to their or our evil or goodness? Machines and technique are *not* autonomous. The laws or principles for technique do not originate in machines, though in their implementation these principles can never be realized apart from machines. People make machines. We operate machines. Humans repair machines. Adults critique and affirm the place of technique in our lives. In short, we are responsible for the place, meaning, and impact used to implement technique. Technique is not autonomous!

Pessimists attribute autonomy to technique because they believe the myth created by the optimistic secular technologist. Confessing ourselves to be law unto ourselves, we think we see this autonomy all around us because this article of faith is thought to be so crucial to our existence. This "seeing" of autonomy by the secularist is parallel to the "seeing" of theists in that sight is related to belief. Just as the theist sees God's handiwork in the design of nature, the regularities of the universe, the miracle of birth, and the renewal brought about by salvation, so the pessimist "sees" autonomy or, to use Ellul's phrase, "freedom" in "self-regulating," "self-augmenting" technique. Freedom or exaggerated autonomy is thus raised to an exalted level, its dictates commanding technique that is thought to dominate life. It becomes the chief ethical imperative for escaping technique's all-encompassing world view.

Freedom or autonomy is believed to be the only sufficiently authoritative force capable of giving life to technologically dominated and, therefore, dead society. Technological necessity demands, because of its totalitarian grip on society, an equally forceful counterpoint that Ellul calls "freedom." Should not the gulag of technique be dialectically opposed by the heavenly hope of freedom? Freedom, because of its exaggerated, autonomous character, is thought to be the only force sufficient to resist technological necessity. This dialectical contradiction between Scylla of technological necessity and the Charybdis of human "freedom" originates in the secular and optimistic exultation of human autonomy, a definition Ellul ironically accepts.

The irony of Ellul adopting the optimistic secular notion of autonomy for his definition of freedom should be noted. By accepting the exaggerated notion of autonomy or freedom, Ellul becomes deeply influenced by the secular optimistic technologist. Accordingly, instrumental or technical autonomous rationality seeks to remake the world after its own dictates. No boundaries or external laws are permitted to interfere with the march of rationality. This march, or cultural progress, is inevitable and leads to total human betterment. Freedom or autonomy is central for this view. Accepting this notion of autonomy, Ellul has chosen the same secular, exaggerated core or religious starting point. Therefore, he does not bring a **radical** critique to technique as much as he does bring a deeply synthetic accommodation to the core virtue of modern technique, an accommodation Ellul wants to avoid. Believing the myth of autonomy, he can only warn us of technique's imposition. He cannot surgically remove the cancer of technique's imposition because he has not penetrated to the root melanoma: the pretension of autonomy. We pretend—literally make believe—we are autonomous, so the ideology will continue, but as I will argue in a moment, life is not at root autonomous.

His dialectical methodology, I must respectfully argue, is not at root, therefore, Christian.[74] His work does not represent an effective antidote to modern technique. Its roots are in secularity; its tensions provide no ease from the burden of technique. God is the Author of all truth. All truth—life's manifold imperatives—is related solely to God. No one imperative—such as freedom—can receive substantive exaggeration without other areas of life becoming impoverished or exaggerated, again a condition Ellul wants to avoid. By exaggerating necessity, he creates an equally exaggerated need for freedom. At the same time, an equally exaggerated notion of autonomous freedom becomes the only accepted antidote to the determinations of necessity. Thus, Ellul must magnify the god of human freedom to compensate for the titan of technical necessity. Therefore, freedom and necessity have become intertwined in a cycle of obsessive mutual need that can only lead to despair.

Ellul's arguments become all the more tragic when one stops to think that he has not experienced technique as a threat in all cases. Technique has brought us a degree of freedom. The mass production and distribution of his books, as well as all of the techniques of writing (which Ellul has mastered so well), have made many of us—including Ellul—freer from the grip of technique by raising our redemptive awareness of the problem. In short, his own writings testify to the reality that we do not live in a technological society. At least Ellul and, I suspect, many others, are freer because of his work.

Neither freedom nor determinism, it seems to me, is the mark of humanness. Responsibility is. Humanness is a response to a variety of God-given norms for and in life, only one of which is the demands of technique. The reality of this statement becomes clearer when one considers that the principles by which humans create and maintain machines are not autonomous products of the mind. Neither my mind nor its thoughts represent the origin of the world. Choice depends on a variety of factors such as genes, social conditioning, class and gender interests, and, above all, conditions for life that are given in life. That is, I respond to internal and to external stimuli or "laws"; my response to these laws characterizes my humanity but their origin is not in me. Sometimes the choice for all practical purposes is nil, as in the case of the color of my eyes. Sometimes it is great, as in how and why this book is written. Response to God, the Author of all life, is always present.

The recognition of the responsible character of life will enable us to accept the pessimist's critique of the imposition of technique while rejecting the flaw of exaggeration in pessimism. That technology comes to us as distorted seems clear. That technique, as a fact and force for life, is inherently totalitarian is denied by Ellul's own writing.

Does autonomy define our essential nature? I think not. A new definition of freedom should be substituted for autonomy. If freedom can be viewed as a response to God's providential and sovereign care for all of reality, then we will increase our ability to respond if and when our nature becomes free of its idolatry of technology. Further, we increase our ability to respond because we can see many more avenues than technology in which to exercise our humanity—avenues that remain in spite of technique's seeming totalitarian nature. Consequently we can decrease the sovereignty—read necessity—of technique's totalitarian nature. Our true freedom is enhanced because technological necessity is reduced—and this reduction comes about precisely because we are able to respond, successfully and forcefully, to the challenge of modern technique. God's sovereignty simultaneously resists technique's alleged sovereignty and provides the basis for our responsible freedom. God created us to be stewards, caretakers of Her creation and we have the ability *in Christ* to manage all aspects of the creation responsibly.

If my charge of Ellul's accommodation to the secular core notion of autonomy is correct, then he does not have two all-encompassing religious ethics, Marxism and Christianity. Rather, his religious root commitment is to a secular notion of autonomy. Thus, his critique and alternative have become less effectual than otherwise may have been the case because the root idolatry of autonomy is not discerned. Because his core commitment is to the primary principle of autonomy, he can articulate no solution to the dire problems he so forcefully raises through his work.

The same critique can be made of the postmodernists' analysis of modern technology. Marcuse, Roszak, and the others make several telling criticisms. A "regime" of experts does erode our democratic responsibility to wisely use technique. "Objective consciousness" is a myth. We do live in an all-too-one-dimensional a society. Nevertheless, both modernists and postmodernists share the same root commitment and thus will share the same unfruitful fate. In each case it is the principle of autonomy—self-rule—that forms the basis for hope and a new order. Modernity has proven that humanity can never be bright, autonomous, omniscient, or wise enough to set the conditions for our own existence. Modernity's failed technological program and fate, as well as postmodernism's failed shamanistic or neo-Marxist worldview and legitimate but short-lived criticism of modernity's failed optimistic paradise, stand as powerful testimonies to the false trust in autonomy.

Postmodernism has yet another problem. It is not so much pessimistic about the use and structure of technology as it is ambivalent. This is admitted by the editor of the book under study.[75] Postmodernism exaggerates the place and the importance of "consciousness" or human awareness. This problem, in turn, leads at least to a superficial analysis of the conditions and the structure of modern technology because it is consciousness that forms the grid for life. At the extreme, postmodernism is incapable of seeing beyond the supremely autonomous thinking subject whose freedom forms the heart of reality. This is especially true for Roszak. Consequently, while postmodernists want to critique modern technology, they cannot find an intellectual program that can guide the proper use of technology. No place can be found for technology because the place and the importance of human consciousness obscures reality.

Furthermore, Ellul's root notion of autonomy does not lead to dialecticism. It leads to an intellectually schizoid worldview. Two "totality views," ipso facto, cannot occupy the same life. This fact is especially true if life is confessed to be whole as Ellul does.[76] Two mutually exclusive sets of principles divide and do not unite the person. Autonomy versus heteronomy, sociology versus theology, Reason versus revelation, freedom versus necessity, alienation versus reconciliation: this is methodological schizophrenia. No amalgam of different sides of the same coin called "totality views" is possible for *no* set of principles together or separate are total in their command of life. Therefore, Ellul offers no holistic, comprehensive position from which to make a radical, integral critique of modern technique; and critique we must. His religious secular commitment to autonomy, together with his dialectical method, form a schizoid and, therefore, not a holistic view. This lack of holistic integrality plagues postmodernism as well.

Finally, Ellul and the postmodernists think of evil or false consciousness as absolute in the historical area and therefore leave us without any historic hope of overcoming the real evil of technique. True, Ellul speaks of hope, freedom, wholeness, and the kingdom of God, and Roszak touts the shamanistic worldview but never with concrete, historical examples. These lofty words are reserved for the transcendent, suprahistorical realm—or, in Roszak's case, part of a dreamy past. History is ruled by technique. Their works are full of examples of why and how our everyday lives are tyrannized by technique. My critique especially of Ellul's aggrandizement of historic evil first and foremost comes from Ellul's stated belief that evil may not assume the place and significance he allots to it. His own dialectical way of thinking calls for "freedom," "hope," "promise," and "joy." That this prophet stands for these virtues testifies to the reality that we do not live in a technological society. At least Jacques Ellul, and many others, do not live for, unto, and because of technique, though our lives have become too permeated by technique.

The problems outlined in Ellul's work are characteristic to one degree or another of other facets of pessimism as well. Others, like Habermas, have similar flaws. The irony of Ellul's using technique to critique and free us from technique is parallel to Habermas using the fruits or the wealth of technique to create an activistic student class of revolutionaries who supposedly will autonomously throw off the "achievement mentality," the root of which is modern technology. The pessimist, in other words, has at the core of his thought a belief in the myth of autonomy that in its technological form is thought to signal damnation while in its humanistic form is thought to be the herald of salvation. What a dark, conflicted world indeed. Personally, I do not aspire to tension, dialectical thought, or contradictions. Wholeness, integrality, being grounded in Christ in the midst of onto-logical diversity seem to be virtues and a state that could provide for a more holistic and happy and peaceful life.

3
Technological Realism

Pessimists concentrate on the loss of freedom that results from technology. Society would be improved if only the demon of technique were exorcized, they say. Indeed, Jacques Ellul argues that technology represents the significant origin of modern enslavement and hence is evil. This evil can be resisted only through power that comes from beyond history because history and culture are subject to technical control. This theistic freedom forms an equally grand dialectic or contradiction to the slavery brought about by technique. This dialectical conflict lies at the root of Ellul's thought.

On the other hand, optimists argue that technology can redress long-standing human problems. Technical and nontechnical problems alike will recede in the wake of beneficent technology. Technical experts can remedy even problems caused by technology. Society would be best served if increasing amounts of modern technology were applied to all human problems. This position results in a society that is overly technological. It remains to be seen if a more moderate position can be found. It can, and "realism" is its name.

I label this position "realism" because it attempts to point out with scientific precision both the "good" and the "bad" of any given technical innovation. It seems to appeal to common sense that technology has its failures and its successes. For example, we enjoy the good of increased freedom afforded by the use of the automobile. At the same time, we experience gridlock, injury, and pollution. Realists want to talk about the positives and the negatives of technology and thus seem more balanced and rational. Realists will seem practical, pragmatic, mediating, scientific, and less negative than pessimists. At the same time, realists are less naive and seemingly more honest than optimists.

94

Realism has scientific, mathematical, and philosophical roots. To expose these roots, a broad understanding of these disciplines is necessary. Specialists within each discipline have collaborated to construct a methodology or a systematic investigative means to evaluate the consequences of technology. Each method's assumptions for determining good and harm, as well as strategies to contain the latter and promote the former, will be analyzed. I will attempt to understand how and why technological assessment follows from the core definition of realism, and define risk and attempt to see how the realist manages risk. I will then explore the notion of tradeoffs and how this notion necessarily follows that of risk. Finally, I will conclude the discussion with an evaluation of the position and with a brief section on the place within human lives the realist would assign to technology.

Realism contains a worldview—that is, a more or less comprehensive way of viewing reality. Understanding this worldview starts with probing the foundation of realism. This foundation is built upon the answer to at least five key questions:

- What does it mean to be human?
- What is meant by utility and by value?
- How does one harmonize technology with other necessary life needs?
- What is the relationship of Reason, technical policy, and beliefs?
- Can we control all or part of technology?

The core of technological realism is seen in this statement by Raphael Kasper:

> On the one hand, it is hard not to recognize that technology has yielded great benefits. However, on the other hand, it has become evident that not all of the consequences of technological innovation are beneficial and that some are quite serious, and often unexpected hazards. . . . It must be recognized that consideration of *only* the potential benefits of technology can lead to an irreparable harm. However, it must also be noted that consideration of *only* the potential risks of technology can lead to technological stagnation, with the concomitant loss of the opportunity for the improvement of man's life.[1]

Therefore, because technology manifests both good and bad consequences, realists argue that we need some way to *assess* the good and the bad. If there is doubt about the potential effects of a technique, then a precise way to measure the degree of risk or exposure to potential harm must be developed. For example, if while boarding an airplane the pilot told all passengers that with precision and certainty it was determined that the plane had a 0.0002 percent chance of crashing before it reached the destination, would you then travel on that flight? I think most of us would because we likely would conclude that the benefit of time saved by flying exceeded the risk of harm manifest in an airplane crash. Suppose that the pilot told you that there was a 60 percent chance that the plane would crash. Would you then travel on that more risky flight? I think not. In this instance your assessment is made by calculating the increased risk of harm versus the potential time saved. Your conclusion would likely be that the danger posed by the increased risk of harm greatly exceeded the good of time saved through flight. This lack of safety might force one to pay additional money to secure a safer means of transportation. This process of risk assessment and probability of harm and associated costs constitutes one of the crucial methods of assessing technological impacts.

TECHNOLOGICAL ASSESSMENT

Scientists and technologists produce innovations that are not fully proven and hence need to be monitored more than once.[2] Assessment thus needs to occur *before* technology impacts our lives, argues the realist, because the consequences of modern technology can be global.[3] Technological assessment (TA) is the ability to anticipate the impacts—positive and negative—of planned technological innovation by scientific means; that is, by analysis, measuring, extrapolation, and research.[4] Thus, TA is the attempt to measure the concrete risks and harms inherent in a technology. Ascertaining the percentage of risk or harm is a key goal. Risk is assessed so that we, the consumers of technology, may use it in an informed manner. Reducing risk is an important goal of realism. To reduce risk and promote safety costs money, however. Costs are expressed in dollar amounts one would be willing to part with in order to reduce the risk posed by a given technology. How and why did this view of technological assessment arise? How does the process of assessment indicate a worldview?[5]

TA is a scientific procedure and not a philosophical exercise. Its aim is to exact precise empirical information so that more rational decisions about the pragmatic use of technology can be made by an informed citizenry. Its intention is not to entertain questions about the meaning and the place of technique within our lives. This exercise must be left to the humanities. The professionals making the assessments, being realists, give no explicit definition to the meaning or to the place of technology in our lives. It remains to be seen whether or not such views are implied and uncritically accepted.

TA attempts to serve broad social goals. Technical means for safety must be attuned to strengthen democratically agreed upon social ends. TA seeks to state precisely the impact and implications of a given technology for the natural and social environment. TA provides a stable and predictable atmosphere for technical innovation that therefore reduces the unintended consequences of technology to a minimum.[6] Reliable and understandable information about technological risk factors are thereby communicated to the public.[7]

I have said that serious risk attends the use of technology. While this is undoubtedly true, realists have determined that the dominant risks faced in modern life are *not* directly related to technology. The predominate risks we face are caused by improper handling of automobiles and tobacco use, as well as from heart problems usually caused by poor health habits. Thus, argues the realist, modern technology is not the greatest problem we face (a not-so-veiled reply to the pessimists). Rather, our *misuse* of technology and our bodies represents the greatest risk to human health.

While TA is capable of detecting and reducing health risks, a realist would argue that we neither can nor should eliminate completely the threat of risk. It is inherent in the use of technology—even the simplest tool may cause *some* type of injury. Further, reducing all risk is not beneficial because the minor threat posed by, say, hitting our fingers with a hammer forces us to use technology more carefully.

Thus, all risk is not wrong or to be avoided. Personal and technological development is dependant on risk—the chance to win or to lose a better quality of life—argues the realist. There is no risk-free life. Development goes hand-in-hand with the threat of losing that which is valued. We are by nature, says the realist, risk-takers. Reducing *unwanted* risks is the goal of risk analysis.

Technological assessment experts are often confronted more by the public's perceptions of danger than by the realities of danger, claims the realist. The fear of nuclear contamination caused by a meltdown is such a fear. That is, the public often perceives that it is in danger and hence forces scientists and politicians to insure safety long before there is any evidence to support the fear. Realists must therefore account for the psychological side of technological assessment, one that parallels the scientific side. That is, the personal, subjective, often

emotive perception of risk goes hand-in-hand with the publicly verifiable, objective, rational, or scientific way of ascertaining risk.[8] Note
should be made of the objective/subjective sides of perception. That
is, scientific, rational, expert certainty is contrasted with popular,
subjective perceptions. When the public's perceptions flare, scientific
realists argue that cool, calculating science should sooth the fears.

TA has a fascinating political history. A Senate subcommittee entitled Science, Research, and Development petitioned Congress in 1965
to create an evaluative agency for Congress. The stated task was to
assess the impact of technology on society. Consequently, the Congressional Office of Technology Assessment (or OTA) was born in
1967. OTA's job was to collect "unbiased" information on the potential or real impacts upon society of all proposed major technological
projects. Reports were then to be made to Congress. This information
was then to inform the enactment of laws.

Typical evaluative questions posed by OTA have been the following. How significant are the "costs" and the "benefits" of the proposed technology? How likely is it that a given technology will
become a public concern in the future? Does the technology have potentially irreversible negative side-effects to its deployment? Can we
limit the technology under consideration? How much time do we
have before this technology will be guaranteed to be safe? Most importantly, what will be the economic and social costs if a technology
causes harm? Thus, an evaluative arm of the Congress was created to
cope with the immense complexity of and occasional harmful consequences of modern technology.[9] No such extensive effort would be required if technology were thought to have no dire consequences.

The term *risk* needs to be clarified further. Risk is the calculated
probability that loss or harm will occur in a specific situation. The
risk of injury may be real, as would be the case if someone dropped a
bomb next to where you were standing. Risk may be caused by fear,
not reality, as would be the case if one feared a military conquest of
the United States in 1997.[10] In the former example, the reality of destruction is certain because an exploded bomb dropped near one's
body is calculated to insure harm. In the latter case, fear greatly exceeds reality.

The risk analyst is to determine the *exact* danger of risk. Thus,
doing risk analysis involves first collecting and then examining the
evidence surrounding any perceived hazard. This process considers
the qualitative and the quantitative sides of health effects and policy
alternatives. Policy is aimed at improving the quality of life. Quality

of life issues tell us how well we are living. Quantitative issues usually involve calculations about monetary costs for maintaining our desired quality of life. Once the qualitative and quantitative issues are presented clearly, then public debate may begin, contends the realist. Does the omnipresent quantitative side of risk analysis predominate and hence subtly but profoundly define qualitative assessment? I will address this question in the final section.

Since costs are shouldered by the public, and since we live in a democracy, an informed citizenry must be willing to take part in a public debate about proposed technologies. This is especially true for technology that may influence public areas such as the natural environment or public commodities such as our food supply. Our government must protect these areas because they are public, or common to all. Only the public, through its governmental agencies, can determine the acceptability of any public risk level. Realists only want to inform this decision-making process.

Returning to the discussion of risk, I find that past research on risk has tended to focus on the marking of hazard occurrence according to location, magnitude, and periodicity. Next, the underlying physical and human causes of hazards were determined. Finally, strategies for managing or reducing potential or real hazards were developed. This kind of methodical but unspectacular work comprises the day-to-day work of many risk assessors.

However, specific, highly glamorous but low-risk threats to our safety too often have captured public attention because of hysterics or theatrics. Many realists argue that risks posed by floods, earthquakes, or technological hazards such as nuclear waste spill or power-plant meltdown pose only minor risks when compared to the risk engendered by smoking two packs of cigarettes a day for many years.[11] That being the case, a realist asks: why do we spend so many of our limited dollars to lower the risk of being harmed by nuclear waste and not on eliminating cigarette smoking from our society?

Risk analysis involves another major distinction. Risks or threats may be voluntarily or involuntarily assumed. That is, you willingly or voluntarily expose yourself to risk when you smoke two packs of cigarettes a day for twenty years. It is voluntary because no external force coerces a decision; motivation comes from within. However, you are unwillingly exposed to risks when you lack firsthand knowledge of immanent danger. If harm is done without firsthand knowledge, then the legal burden often shifts away from the injured party. In the former case, public sentiment and law have tended to hold the person responsible for the risk, although this notion has dramatically changed because of the tobacco companies' settlements. In the latter case, sentiment and law tend to fix responsibility on external agency.

The voluntary/involuntary distinction becomes clearer when you look at the risk posed by hazardous jobs. Some occupations are more hazardous than others. Thus, they are awarded more pay. High-rise construction workers are paid more because their jobs expose them to a greater degree of physical harm than do many jobs. Moreover, these workers also must be concerned about the public's safety. If something as small as a penny is dropped from a skyscraper and hits one, then injury may occur. Initially, public sentiment supported highly compensated workers because they were exposed to increased risk.

Recently, however, many are beginning to question the equity of this arrangement. Questions surrounding the ethical legitimacy of exchanging money for health and safety, the possibility of a lack of worker awareness of the degree of risk, and the efficacy of the extent of legal protection for endangered workers make this debate heated.[12]

There are several related parts to the ethical and legal context of risk analysis. There is the voluntary/involuntary distinction just mentioned. There is also the question surrounding the familiarity or unfamiliarity of risk. If the risk is familiar, then the threat legally and ethically is said to be less onerous and hence less weight is attached to a harm suffered. The legal and ethical risk to the company would be minimal if a motorist crashed because a well-worn tire blew out when traveling at high speeds. The tire company would suffer little, if any, legal pressure because the driver's lack of tire maintenance increased the risk.

Moreover, a lower probability of risk usually signals a reduced expectation of harm. Returning to my motoring example, I may say that the ethical and legal onus rests on the shoulders of the tire company and the dealer if a tire on my auto blows out on the way home from the showroom. This is so because the low possibility of harm was assumed and guaranteed by the tire manufacturer.

The time horizon also plays a role in consideration of risk. Does the risk lead to immediate or long-term harm? Risk that is relatively immediate—thermonuclear destruction during the Cuban Missile Crisis—seems to weigh much heavier on us than do future threats such as exposure to leaking nuclear wastes.[13] Short-term risks often arouse us to action more than do long-term risks. Long-range risks such as those posed by years of cigarette smoking are more difficult to litigate successfully until incriminating documents exposing tobacco executives' knowledge of the danger of cigarette smoking are released.[14]

I have shown the consequences of risk and harms of differing import. Now I move on to consider the consequences of different kinds of risks. Consideration of these categories will add further scientific precision to this analysis. There is a type of risk that is high in probability and high in consequence. For example, there is a high chance that if one were in a head-on collision going sixty-five miles per hour, one would likely pay a high consequence or receive serious harm. Conversely, there are also low-probability, low-consequence harms. Suppose a dam were to collapse (only one in three hundred suffers some kind of failure) you live on a high hill several miles from the dam. If the dam collapsed, there is a low probability of damage to your property and a low consequence of harm: the probability is reduced because of distance and height of the hill. Identifying consequences is extended to cover incidences of low-probability/ high-consequence harm (nuclear exchange with the former Soviet Union), or a high-probability/low-consequence harm (the risk of eye damage from viewing color television). Risks are calculated in consequential terms. Neither knowing the consequences nor acting on the data can eliminate the uncertainty that goes with the use of technology.

Technology, just like life, always involves an element of uncertainty. One can never be absolutely sure of the outcome of a given event; one can only calculate the probability of an outcome in an inherently uncertain situation. Probabilistic risk assessment attempts to determine, with mathematical precision, the probabilities of event occurrences. If an event occurs often enough, such as my car starting when I turn the key in the ignition, then I can predict with some confidence the outcome of similar future events and thus reduce my uncertainty and increase my certainty in my use of technology. All of this depends on having a sufficient number of occurrences to warrant honest estimates of risk. The goal is scientific precision with increased certainty in the face of probability. Relative certainty is attained through securing objective, rational, precise scientific information. Through methodological analysis of factors of risk, probabilities are determined. This kind of science is the first form of reason noted. These results are supposed to be value-free, open to universal investigation, and applicable to all people. Insofar as they are thought to be value-free, they are thought to be the opposite kind of sentiments from beliefs. In fact, contends William Rowe, beliefs of all kinds play a much larger role in investigation than previously thought.[15] That is, while estimations of probabilities necessarily involve all the methods of science, these methods themselves assume basic certainties that are not publicly corroborated or verifiable. Core beliefs such as the ideal of progress, the religious-like trust in Reason, and the quest for certainty through science are among the beliefs Rowe has in mind. The nature and the function of these shadowy but deeply influential dogmas will become clearer shortly. For now I need merely note that core beliefs are involved in risk analysis and in probability studies. We may conclude that Realists believe that technological reality presents us with an inherently risky situation that can only be made less threatening through the surety brought about by science. That is, technology carries with its inception *inherent* benefits and *inherent* risks of harm. Only science and its methods can deliver us to a more secure future. Thus, there is a need for scientific control to overcome a risky situation.

RISK REDUCTION

There are two basic approaches to risk reduction. They are prevention and limitation. The first means to lower the probability of the unwanted event. The second means to render the consequences less unpleasant when it happens. Storm cellars for tornadoes are purely mitigative; no one can steer tornadoes. Cloud seeding, in the hope of changing the meteorological conditions that lead to the tornadoes, would be preventive if it worked.[16]

Risk reduction involves an aggressive policy whereby an attempt is made to mitigate the unwanted effects of technology both physically and psychologically. Thinking—assessing risks—is complemented by doing—managing risks.

Law provides the predominant deterrent to risk increase. Regulatory agencies of various levels of government attempt to make our lives safer through passing legislation and enforcing its sanctions. Before a measure aimed at reducing a risk becomes law, however, it has to become a public or at least a powerful private concern. Before a concern becomes an agenda item on a legislator's docket, sufficient numbers of the public or sufficiently powerful private interests must become alarmed enough to push for a judicial hearing. Once sufficient concern is registered, legislative hearings begin to determine if more people, especially experts in the field(s), believe there is a problem posed by a given technology. Debate is then initiated as to the degree and the nature of the problem. A verdict is rendered, with an action supposed to follow.[17]

It is important to note that even if (and this is a large if) the government enacts a law, it does not always mean that the law stipulates the exact level of risk or the degree of safety necessary. A degree of flexibility in interpreting the law and setting proper standards is left to the regulatory agencies.

Moreover, determining the exact level at which risk occurs is impossible at times. Assessing and reducing hidden risk often requires draining depleted reserves for minimal results. An example will clarify this point. If the public wants the risk factor of auto fatalities reduced by 5 percent, then we may have to pay an additional $275, say, for anti-lock brakes. If we want to reduce our risk of auto fatalities an additional 7 percent then, we may have to pay an extra $250 for airbags. Thus, risk reduction—a benefit—costs money, sometimes more than we are able to pay. Or in the case of airbags, the protection afforded may greatly exceed economic costs, as well as the threat posed to young children if hit by an airbag.

The problems with laws requiring risk reduction are not only the increased costs but the vagueness of the laws that rarely stipulate the exact level of acceptable risk. This process of determining acceptable risk is left to the democratic process, which in turn depends on our collective ability to pay. Given today's almost universal tightening of resources, a careful, well-thought-through plan of risk reduction is mandatory; so argues the cost-conscious realist. (I will return to this careful stewarding of monies in the section on evaluation.)

H. W. Lewis portrays this problem of payment and vagueness about explicit levels of acceptable risk reduction perhaps in a clearer manner.

> The issue of determining a proper level for regulation, given the costs and the benefits, is the most perplexing problem. A license to operate a nuclear power plant is issued after a certification that the power plant can be operated 'without undue risk to the health and safety of the public.' That is the standard. No one has ever defined how much risk is due the public, and the question, when raised, is normally met with an embarrassed silence. This theme pervades all regulation.[18]

It should be clear that there is no risk-free world, and if we want to reduce any given risk we must stipulate how much we are willing to pay for exactly how much risk reduction. We can reduce or counterpoise risk with other goods, evaluate, and then manage it. We cannot eradicate it. However, when risk can be lowered to a finite level such that it is considered negligible, realists' speak of this level of risk as *De Minimis* or minimal risk. There is an extremely small chance that the chair in which I am now sitting will fail with the consequences that I will be seriously injured. Therefore, based on past experience of sitting in this chair, I may conclude that the chances of my getting hurt because of defects in this chair are 1 in 1,000,000 or *De Minimis*: negligible. More money spent in reducing the risk even further would be money poorly spent.

While the chair I am currently sitting in may be safe, not all technologies are this safe. Let us assume that I want to reduce my risk of serious auto fatalities, yet would be financially pressed to spend an additional $250 for airbags. In that case, I must weigh the risks of auto fatalities and decide if the peace of mind afforded by airbags is dear enough to forego other needs or pleasures. Should I decide to give up other needs/pleasures for the airbag, I have made a *tradeoff*. A tradeoff means that I reduce the quantity and/or quality of one good or need to increase the good or need of another, more valuable, need or good. In this case, I have decided to give up going to the movies for several months so that I may be safer in my auto. That is, risk analysis demands that we ask ourselves what we are willing to trade or give up in order to increase the certainty that technology will be less threatening.

Many, if not most, tradeoffs involve calculating the balance between product value/cost/durability and efficiency of purchase and maintenance. The calculation of the enjoyment gained by use of one object over the enjoyment use of that of another is called *utility calculation* and its substitution called *utility tradeoff*. This calculation of utility tradeoffs is necessary to understand if we are to discern realism's ability to define not only happiness but worth as well.[19] Thus, the notion of utility, which means value *and* happiness, is important to note. I will return to this theme of utility several times throughout the chapter.

The Delaney Clause is a legal effort at risk reduction by attempting to monitor food additives. In 1958 Congress set out to counteract the risk of contracting cancer by enacting into law what came to be known as the Delaney Clause, Section 409 (c) (3)(A), sponsored by James Delaney of the House of Representatives, to the Food, Drug, and Cosmetic Act. The act states that the Food and Drug Administration will allow no food additive to be placed in food or added to drugs if it is known to be carcinogenic in animals or in humans.[20]

The wording of the Delaney Clause is interesting. It states, "No additive will be deemed safe if it is found to induce cancer when ingested by animal or by man, or if it is found, after tests which are appropriate for the evaluation of the safety of food additives, to induce cancer in man or in animal."[21] Perhaps the two most apparent aspects of the bill are its simplicity and its rigor. Negligible risk will not be tolerated. Absolute zero risk must be the standard.[22] The simplicity—zero risk—is the standard; the rigor is what follows. Ridding our diet and life of any food additive that is known to constitute even *minimal* risk is both extremely difficult and quite expensive, argues the realist. Why not set the upper limits of risk at, say, one in a million chance of harm from additives?[23]

One may wonder why the Delaney Clause is so rigorous in its regulation of carcinogens. To address this question and to give an example of how the Delaney Clause influences risk analysis, I look at chemical carcinogens. The effects of build-up are insidious and cumulative. The health risk is negligible during the build-up period. Over the course of a lifetime, however, the cumulative effects of carcinogens are especially dangerous and often fatal. Therefore, the banning of chemicals from our food is of the utmost importance. There are approximately 65,000 chemicals listed. Of these, 150 are known to be carcinogens.[24] Therefore, the need for strict monitoring is mandatory.

The case of the sweetener saccharin represented a focused test for the Delaney Clause. In 1879 saccharin was discovered by a chemist at Johns Hopkins who was practicing a technique to oxidize some toluene compounds. One of the unintended consequences of the experiment was a sweet powdery substance that covered much of his cloths. The distillate was discovered to be approximately four hundred times more sweet than regular sugar, while containing fewer calories.

The artificial sweetener waxed and waned as a safe favorite for Americans until the 1970s. By then, increasingly sophisticated monitoring devices suggested that the use of saccharin put the public's safety at risk. Laboratory evidence suggested that this sweetener could cause bladder cancer in small animals. However—and this is crucial—the evidence for saccharin causing cancer in humans was much weaker.

Battles ensued. The realists took the saccharin makers to court prosecuting under the Delaney Clause. The outcome was historic.

Essentially, experimenters found out that a methodological presuppo-
sition had unconsciously influenced this entire project of risk analysis.
The uncritically accepted scientific assumption was that the risk
faced by the animal is the same as that posed to humans. This assumption is based on the belief that an evolutionary genetic link between the animal and the human overrides genetic dissimilarities.
Secondly, scientists also found out that no bladder cancer is seen in
the subject mice or rats until their lifetime diet consists of approximately 1 to 2 percent of body weight intake of saccharin per test
period. Only after long-range exposure at high doses do a few animals
develop tumors. Cancer risks do rise but only with higher concentra-
tions of the sweetener. A 1 percent saccharin diet for mice would be
the equivalent in people to a consumption of about a quarter of a
pound of sweetener per day. This would in turn correspond in sweet-
ness to about a hundred pounds of sugar a day.[25]

The outcome of a lengthy court battle has come to a virtual stand-
still. Saccharin has stayed on the market, but the risk of cancer is not
absolute zero as the Delaney Clause demands. The risk is about ten
chances in one million that one will get bladder cancer from two to
three packets per day saccharin consumption. Thus, we have chosen
to continue to use saccharin because the benefits—increased sweet-
ness and decreased calorie consumption—outweigh the potential negatives—bladder cancer—from this product.

It is clear that technological assessment and its more specific relative risk assessment require a great deal of scientific and technical expertise. The implementation of many technologies is sustained
because only a select few with specialized knowledge can assess technology. Thus, assessment of modern, more complex technology often
escapes popular scrutiny. The public—that is, people with little or no
training in a given field—is deemed ignorant of modern technical
complexities or do not want to be bothered by evaluation. Hence,
much of contemporary testimony before legislators is done by experts.[26] In fact, the public increasingly depends on experts for reliable
testimony and information. The consequences of this trust in experts
will become plain by the end of this chapter.

While experts may have the knowledge to analyze the risk present
for a given technology, it is worth asking whether they are competent to determine the "benefits" of a technology. Can any expert tell
other humans what is good for them? If we say no to any degree to
this question, we must develop a method of determining public
good.[27]

The expertise of the engineer often is tapped to determine public good. Virtually all technologies are engineered. Engineers are trained to construct projects safely, usually with margins of safety built into the specifications. The safety margin is the scientific attempt to create a project, find its capabilities, and build in an additional somewhat arbitrary margin where the risk of failure and subsequent injury is reduced. However, the extra time and material used to construct safer artifacts involve tradeoffs.[28] Additional safety involves higher costs and more time to construct the product. Moreover, the engineering expertise requires focused technical skills plus increasingly broad public relations skills associated with testimony before the courts and Congress. That is, engineers are called on to justify the safety of their projects to various judicial and legislative agencies.

The government is the largest, most influential institutional user of technological information. Data are collected to ascertain risk and to determine the level of regulation necessary. Much of the government's involvement in technological assessment and regulation came about as a result of the increased capacity of modern science and the resultant concern for our safety. One can see governmental realism in this classic realistic statement from John F. Kennedy. He specifically mentioned,

> our responsibility to control the effects of our own scientific experimentation. For as science investigates the natural environment, it also modifies it—and the modification may have incalculable consequences, for evil as well as for good. . . . The government has the clear responsibility to weigh the importance of large-scale experiments and to advance the knowledge of national security against the possibility of adverse and destructive effects. . . .

While the former president started on a realistic note, he ended on a more optimistic one.

> As we begin to master the destructive potentialities of modern science, we move toward a new era in which science can fulfill its creative promise and help bring into existence the happiest society the world has ever known.[29]

Sadly, neither the resultant history of technological deployment nor government assessment of technology have justified this optimism. The formation of a national technology policy and the setting of priorities and timetables for the deployment of major technological artifacts has been sporadic at best. In the last thirty years, the hazards and the problems associated with technology continue to mount. The safe disposal of nuclear waste is a case in point. Thus, the concept of control, contained in both of the above quotes, must be questioned.

The government established the Office of Science and Technology (OST) and related subcommittees to serve the president.[30] OST's job is to coordinate the various governmental research and development projects, advise the Congress and the president, and provide a means for OST to lobby Congress for its interests.

The following quote, betraying both the spirit and the work of OST, defines Congress' relationship to the technical community in particular and to the positives and negatives of science and technology in general. Note should be made of the spirit of realism.

> Our experience to date is that the Congress does need *advice* and that judgements on technical matters must continue to be rendered within the conventional legislative process. . . . What many members of Congress have become is sensitive to the capabilities and the limitations of science and technology.[31]

Thus, the spirit of realism wafts throughout governmental policy.

The process I have described thus far is standard fare for our democracy. It should not be concluded, however, that this advisory process functions smoothly at all times. Conflicts often emerge during legislative hearings. Conflicts can produce greater public support if the roots of conflicts are resolved. Harold Green portrays this adversarial spirit and its impact on the assessment of technology:

> When I speak of an adversary process I am not suggesting a formal adjudicatory process in which opposing parties contend through legal mouthpieces. . . . Rather, I am suggesting only that a mechanism be developed that will permit and facilitate the articulation of public policy of all relevant facts, pro and con. And when I speak of technology assessment, I do not encompass the assessment of the full potential range of the social consequences of a technology. Rather. . . I am concerned with technology assessment solely from the standpoint of identifying and controlling those attributes of a technology that adversely affect basic human rights [such as] . . . the health, safety, and security of the public.[32]

This quote is prompted because much of the optimistic hype for a pending technology, usually under consideration for Congressional funding, has proven to be vacuous. The benefits of technology are pressed while the potential negatives and risk factors are hidden. Seeing this, realists spring into action with lists of potential problems. A debate predictably follows that challenges interests.

Optimists conceal risks for a variety of reasons, thus causing conflict with realists. The economic pressures of capital investment, research dollars, and overhead force optimistic lobbyists to exaggerate the potentials of technologies. The special interest group for the "atoms for peace" nuclear program is an outstanding example of this pressure. Secondly, risks and problems are often unplanned. Human intelligence is limited and thus its projections are fallible, a fact not fully appreciated by optimistic lobbyists. Finally, advocates of a particular government sponsored technology are better funded usually than are private groups more critical of a proposed technology.

Thus, adversarial relationships between proponents and critics of technology have been played out on a less than even playing field. Funding, information, legal aid, and public relations expertise typically have sided with technological enthusiasts. This in spite of congressional attempts to mitigate the problem through the creation of OST. Realists have tried to remedy this imbalance.

Congress wants to protect OST from being tainted by lobbyists' money by assuring a more pluralistic, less money-driven view of technological assessment. Therefore, OST includes such relevant communities as engineers, environmental consultants, consumer watchdogs, related governmental groups, and labor representatives.

Realists agree that it is necessary to involve the president of the United States in the development of a technology policy. The president needs to articulate a national technology policy to Congress and the American people. The influence of the president that is now implicit needs to be made nationally explicit. Realists argue that if the president were more explicitly involved, then citizens would become more involved. If the president proposed a technology policy that articulated the relationship of, say, technology to environmental concerns, then a debate would result as conflicting interests surfaced. We need democratic explicitness. We now have a hidden, implicit, undefined national policy. When Congressional debate does occur, projects such as the Stealth bomber are subject to the influence of behind-the-scenes lobbyists. Open national debate may influence policy that may change the kinds of appointments the president makes. The appointment to the vice presidency of Senator Al Gore, in the context of his popular book on environmental degradation, is an example of how the national debate focused on how concerns about the environment influence appointments. In all of this and more, the realist argues that assessment must complement a national technology policy.

It would be a mistake to assume that the assessment of risk is an objective, rational, scientific process. Risk assessment is not an objective scientific project for three key reasons. First, risk involves social and personal perception. Perception is the key word. Dramatic accidents, realists might argue, have caused an overestimation of certain risks. The nuclear power industry is an example of this problem. Using nuclear power is much safer for a random sample of the population than is smoking two packs of cigarettes per day for many years. The public has become alarmed because of accidents, such as the near meltdown at Three Mile Island. Consequently, we have demanded more safety from the Nuclear Regulatory Agency than we have, until recently, from tobacco firms because our historic perceptions of the risks posed by nuclear energy is that nuclear power has posed the greater threat.

Moreover, when arguing for or against a controversial technology, the perceived trustworthiness of the person is crucial to the public's perception of risk, argues the realist. If we tend not to trust a person or a community, as for example the lack of trust extended to the "environmental" community by the "technical" community, then we will tend to minimize its evidence. Perhaps this difference is due to competing and differing ideologies. Perhaps it is due to conflicting economic interests. One factor is certain: community, institutional, and personal values and commitments influence how the public perceives risk.[33]

Thus, we may say that risk assessment is not a value-free, unbiased, rational, ethically neutral process. The perception of risk contains a social process. Perception, bias, class interests, and commitments enter into the perception of risk *for the public*.

The scientific methodologies comprising evaluation are value-free, argues the "naive" realist.[34] Further, ethical and philosophical values do not taint the work of the scientist. The thought of naive realists leads us to conclude that all realists are not subjectivists.[35] Subjectivists are in fact the historic minority. Naive realism represents the majority opinion. Naive realists, supposing a more reason-centered, value-free view within realism, sharply draw the line between subjective and objective interpretation.

According to this view, analysis is done in separate stages—those I have delineated. The first stage of investigation through the last stage of drawing hypotheses are conducted in an allegedly rational, objective, value-free manner. I call these naive realists *objectivists* because they conclude that the entire data calculating process is value-free and, therefore, objective. The final stage of analysis is thought to produce publicly verifiable, ideologically neutral objective facts. Only after analysis is subjectivity thought to enter. Only at this time *should* we start talking about policy alternatives and ramifications. Political forums are the places for partisan opinions—so naive realism believes.

We must understand that the "naïve Realists" are dependent upon a Positivistic philosophy for their moorings. That is, reality and scientific interpretation are believed to be value-neutral according to the Positivists. There are no principles, no bias, and no taint of ideology inherent in reality or in thought. Others are ideological; we are not argues the Positivists.

I insert this objectivist/subjectivist debate for several reasons. First, I am arguing that debate about the place of technology within our lives and the principles that guide our assessment *should* take place in a public, that is, a **pluralistic forum**. The term "public" does not denote an ideologically undifferentiated community. Many ideologies combine to form the "public." Second, the debate raises the question as to *where* the authority for ethical norms for risk assessment should be located. Both positions fall back on a foundation of certainty used to cope with the uncertainty inherent in technique. The subjectivists place certainty in the social context or public/political debate. This is a "democratic" solution. On the other hand, the objectivists or naive realists place their faith in human reason as it is manifest in scientific expertise.[36] This debate will be addressed in the concluding section.

It is important to outline clearly the several elements that are part of this debate. First, there are the scientists and their data. Then there are the public and the social context they represent. Third, there is the government and the legislative and judicial processes. Last, care must be taken to see the various grids—worldviews, ethical principles, points of bias—that exist in the minds of the scientists, public, and the legislators. These four major categories, in turn, can be further subdivided so that the end result is a complex labyrinth of factors that affect risk assessment. While the debate rages as to *how* the process should be conceptualized, there is no debate as to *why* the process should continue: because technology presents us with significant opportunities for both harm and benefit.

Specific ethical principles habitually temper human analysis. Some realists would acknowledge these principles. Community, social and professional interests, beauty, freedom, sensation, geography (to name but a few) influence our assessments. We generally spring to action when our families and our neighborhoods—two primary communities—are threatened by technological uncertainty. If the beauty of a sunset is threatened by auto exhaust, then we generally become concerned about the risk because the beauty of a sunset is believed to be priceless. Cherished ideals like freedom of movement are cultivated, especially in Western democracies, and thus affect our assessments. In the final section I will return to the influence of principles—or, to use language previously expressed, "hidden values"—and their part in realism's program.

I do not intend to give the impression that in any given debate about risk the public is energetically involved in an adversarial relationship with scientists. The public often tends to minimize risk for many reasons. Cities depending on one primary industry often find it difficult to implicate the area's primary business. I experienced this denial while growing up in a steel town in western Pennsylvania. Residents found it difficult to concede that smoke emissions from the factories constituted a health hazard. Many of us blamed paint manufacturers for producing poor quality paint when our house paint prematurely chipped. Only after much research, debate, and rancor were we able to admit that the problem originated in smokestacks.

Moreover, research indicates that as personal satisfaction increases, as measured by levels of employment, income distribution, and quality of life indicators, the perception of risk tends to drop. That is, as indicators of personal well-being rise or at least are perceived to be sufficient, one tends to minimize the threat of a hazard. If the threat is publicly perceived to warrant action, however, a homogeneous community will often tend to trust local officials to cope technically with the problem.[37]

The common factor in these two examples is the fear of losing money and the perceived security that comes with money and technology. A fear of economic loss increases resistance to acting on perceived risk. This dynamic is especially the case when the threat is reported by those external to the secure community. If outside experts are telling a specific community that the local steel mill is a carcinogenic threat to their lungs, and the company counters by saying that installing "scrubbers" to remove smoke will cost jobs, one is not inclined to trust an outside expert's risk estimates, especially if one's job is threatened.

Further, if one's family is independent and sufficient assets are located beyond the threatened area and if retirement is at hand, then one is statistically more apt to minimize risk. The perception of danger is lessened because one's fear of loss has been reduced. Comfort becomes a greater need than does activism, at least until the threats become more immediate.[38]

Finally, there is a geographic bias against premature risk prevention. If one is deeply rooted, then one hesitates to believe that a technological danger exists if local authorities say it isn't so. Authorities would never cover up problems, the belief holds, because one personally knows the authorities' character and expertise and can therefore trust these people to do a credible job. Thus, there exists an inherent conservatism—an inability to question the status quo—because of vested interests, attachments, economic and class advantage, and ideological bias. These phenomena help explain the initial lethargy of Love Canal region members when first told of the dumping of chemicals into the canal by Hooker Chemical. Emotional and economic ties to the company were especially strong determiners of initial lethargy and hence of risk perception.[39]

RISK MANAGEMENT

We have spoken about the assessment and the reduction of risk. Now we turn to the management of risk. This section will discuss the historical forces that caused risk management to emerge. It remains to be seen if the spirit of realism—the discernment of the positives and negatives of technology—holds for the management of technology.

Risk management is the regulation and control of the technology that we create. The predominant modern impulse leading to the attempt to manage technology was the control motif in science as outlined in the chapter on optimism. It must be remembered that the management of nature meant the control of its products. The cry for control of technology was heightened by the Industrial Revolution. The size, scale, and frequency of accidents and injury resulting from the Industrial Revolution increased the need for protection from technology. The number of exploding boilers, for example, between 1816 and 1848 to which the federal government had to respond proved to be the beginning of modern risk management.[40] Laws were enacted, inspections were initiated, scientific testing increased, and consumers prodded the market to produce a safer boiler.

Why has risk management arrived so relatively late on the historical scene? Part of the answer is located in the growth of the scale of technology and its related problems. Prior to the Industrial Revolution, technology did not pose a significant threat because its size and capacity were limited by social and ideological factors. Additionally, technological optimism prevented us from seeing the serious and systemic ills caused by technology. Scientists and technologists were thought to be incapable of any *technical* wrong, so the belief in progress held. The first modern scientific society—the Royal Scientific Society—argued to the king of England in its charter that an investment in science is a stake in human prosperity.[41]

Furthermore, optimism's belief in the moral and technical goodness of humanity continued in spite of the horrors of the Industrial Revolution. Technology was producing, by the end of the nineteenth century, more goods and providing the wealth for more services than any technology in history. Industrialism led us to believe we overcame the long-standing human dilemma of want. Additionally, our form of government stressed personal freedom—freedom of contract and individual initiative. This made Americans loath to interfere with its burgeoning industrial prowess because it was built on individual initiative; so the myth went. Concomitantly, the nineteenth century saw the emergence and the increased popularity of a theological movement know as "perfectionism" and its parent revivalism. Both Protestant Evangelical movements shared the common American assumption of the inherent goodness of both humans and our social and technical support systems.[42] When Protestantism and liberal secular democratic forces became the hegemonic power in this country, any more realistic assessment of technology was going to have to wait for the horrors of World War II.

Transition

I have argued that technological realism represents an attempt to understand the "good" and the "harm" of technology, using precise scientific methods and procedures to assess the level of risk posed by a technology. I have shown that there is no risk-free life; indeed, argues the realist, people are by nature risk-takers. Further, TA is not merely a cold, rational, objective, scientific, value-free procedure.

There is a subjective and a social side to the art of determining levels of risk. Central to my analysis is the contention that the uncertainty posed by modern technique has caused realists tacitly to rely on a cluster of principles or values. Finally, I have argued that realism is looking for some explicit, secure basis of certainty.

The imposing nature and complexity of modern technology has forced us to try to manage its outcomes. The ultimate goal of this attempt at control is human happiness. This happier state can be represented by the principle called utility. We trust in or are certain of our rational scientific methods to help us attain happiness.

Finally, I raised the question of the place of the expert. Can the expert alone or primarily be relied on to guide the ship of assessment to the shores of safety? Or, must assessment be done primarily through the democratic process?

THE PLACE OF TECHNOLOGY

Realists allow less room for technology than do optimists but more than pessimists because of one influential concept. Realists are forced to trade off technological realities for nontechnological goods because life consists of a multiplicity of things and activities that have value. Thus, the context or the matrix of life impinges on realists' conclusions so that they are forced to adjust some technology to other life needs. At the same time, realists want to assign technology a place within our lives, contrary to pessimists, because technological goods can enhance and not merely harm life. Realists are more interested in finding a social and a personal context for technology than in imposing more technology on an already overburdened public. Context is the operative word because tradeoffs are always made to accommodate technology to other demands in our lives. Therefore, technology must compete with other necessary aspects and limited resources for its place within our lives. Thus realists ask, "What other life pleasure or need are we willing to forego to increase our use of or control over technique?"

Tradeoffs imply prioritizing. Prioritizing involves values and principles and thus is an ethical discipline. It is not just the economic cost that is being weighed when we calculate our priorities. We are saying we hold dearer more of X and need less of Y because we value X over Y.

Still more is implied in the art of tradeoffs. A view of life is implied in the art of tradeoff. Less or more technology is a way of measuring the *worth* of something or some activity *in relationship* to some other thing or activity. If we want more automobile safety and hence are willing to pay more for autos, then we are saying that our life and safety are worth more than money or other technical objects or activities sacrificed. The money and the inconvenience are viewed in the context of the other goods in life and thus a judgment is made. Thus, the context and the values of life determine our technological priorities. The question is whether realism sufficiently develops a view of a context, or an ontology, so that we can properly decide on tradeoffs. Relatedly, one must ask whether a full or holistic view of the person is present, or whether a myopia exists that restricts the realist's attempt to develop a view of tradeoffs sufficiently. That is, are we trading off commensurate or incommensurate goods? Relatedly, exactly how should the public weigh tradeoffs?

DISCERNING REALISM

What can be said about the strengths and the weaknesses of technological realism? It is almost axiomatic that human beings experience the "good" and the "bad" of technique. We drive cars and experience the freedom afforded by travel. Nevertheless, we also endure gridlock, smog, noise, and even the death of loved ones. We endure risks and harms because the promise of freedom afforded by the auto outweighs our crude calculations of the threat of injury. Therefore, there is a common-sense appeal to realism. This approach seems to provide balance to the one-sided attempts of optimism and pessimism. Balance is a rational virtue because it avoids the moroseness of pessimism and the jubilant, naivete of optimism—but does balance provide a deep enough interpretive principle?

I have often capitalized Reason in this book for reasons I have developed. I now use it again in this chapter for yet another reason. I mean to encapsulate under the term Reason at least three distinct uses of scientific rationality. The first use is that of quantification.

I appreciate the rigorous attempt to quantify the benefits, harms, and costs of technique. By presenting us with the percentage of risk and harm inherent in a project, the risk assessor helps us weigh our technological decisions more carefully. If we say we want nuclear power but don't want nuclear waste in our states, and if we know the percentage of risk associated with radiation leakage, then discussion about the cost of disposal becomes more meaningful. I am arguing that these kinds of calculations could force people to wrestle with the place that technology should occupy within our lives. When we *trade off* or replace a technical good with a nontechnical good, we prioritize—that is, we say that a nontechnical good has worth. Thus, we enrich our lives by at least three ways. Through our calculations and by enriching extra-technical areas of life, we expand our horizons and choices. Then, by limiting the scope of technology, we resist technicism.

The scientific precision of assessment adds needed exactness. We can appreciate the scientific rigor present in methods of measurement, hypothesizing, analyzing, and predicting. These methods can advise us of the justice, stewardship, and environmental and social responsibilities of our technologies. If we determine the more or less exact extent of the risk of nuclear pollution, then act to reduce that percentage, we are using our science to protect that which we value and thus are acting responsibly. Science can and often does strengthen collective responsibility.

This is so for several reasons. We live in an era when budget restraints force us to prioritize our needs. Therefore, if public discussions about tradeoffs lead us to question and to prioritize economic resource allocation for technical safety, then a necessary stewarding of our financial resources has taken place. Realists have begun this conversation. Also, realists partially recognize that there are other aspects of life and that those aspects are not dependent upon nor are they an outgrowth of technology. I label this view of rationality as *instrumental* because technical reason acts as a means to the end of a safer technology and a richer life.

Contrary to the optimists, realists could have a richer view of life. The environment, say, is not just "stuff" on the way to technical absorption. Its chemical, biological, and physical composition require care if we are to keep it as well as ourselves with integrity. I will demonstrate in the next chapter how the structuralists make the context of technology the focus of their discussion, and thus advance our knowledge even more. For now it must be noted that to label a nontechnological thing or an activity a "good" *could be* to see the value of other areas of life if a sufficient view of the context of life were developed. Unfortunately, realism does not do this.

An interdisciplinary methodology is at least implicit in realism and thus represents another use of reason. This methodological principle provides better potential assessment of technology's impact because more of the context of technology is understood. Talk about costs, tradeoffs, and "goods" is or could be couched in a discussion about the context of technology. This talk is important because the context of technology is the matrix of our life. The legal, philosophical, ethical, economic, natural areas of life should be of concern to us because they constitute the arena for living and thus interrelate with technology. Thus, realism raises the quality of discussion by at least implicitly questioning the context of technological decisions.

When the discussion about technology comes to the floor of Congress (or any other "public" forum), an additional strength is *partially* in evidence. To the extent that a plurality of technological communities and positions are represented at such gatherings, justice and responsibility can be furthered. Justice could be furthered because a more diverse group of persons could be heard. Responsibility could be furthered because nontechnically oriented people are forced to assess the effects of technology for themselves and thus reduce their reliance on experts. Public—that is, pluralistic—discussion about risk management thus can enhance responsibility. Unfortunately, again, realism falters because it leads to a heavy reliance on technical experts. Those of us who do not earn our livings in technical areas must shoulder some of this blame for this loss in responsibility.

Understanding how pluralism could work should help us improve our assessment. Pluralism applies to the political heart of risk assessment. Gender, geography, race, class, education level, and political and economic interests affect analysis and implementation. People are not isolated, unconnected individuals. They belong, per their identities, to primary communities like families, schools, nations, neighborhoods, places of worship, and so on. In this plural context, justice means first the official legal recognition, then the preservation and enhancing of the technical and supratechnical needs of these communities. Recognizing this fact would improve subsequent analysis because it would become a more realistic (that is, multifaceted) human endeavor. In this regard, realistic subjectivists have also contributed limitedly to collective wisdom.

I have continually stated that realists could add more depth and richness to their deliberations if they possessed richer views of reality or ontology. A richer view of life is not present because realists manifest no concept of the structure or the context for the diversity of all living and nonliving things. Nor can they relate the structure of technology to the diversity of life because there is no ontological principle of law beyond that of Reason. There is no authority beyond Reason to shape life. Reality and its different aspects are not rooted in any lawful structuration because there is no *principle* in life that makes for identity beyond Reason and its cognates and its ability to connect different facets of life.

Therefore, all they can do is talk about "tradeoffs" in an abstract—that is, uncontextualized—pragmatic manner. There are no inherent lawful principles nor are there structural connections embedded in other modes or areas of life that would lead them to act in any given manner. Thus, there is an inherent **pragmatism** to realism. The phrase "a beautiful sunset" and the principle of beauty contained therein is a reality that exists beyond my concept called beauty. It involves at least the realities of the sun's gaseous explosions, its refracted light, and one's social and psychological state. These realities and their laws impinge upon any deliberations and must therefore be given significant and systematic consideration.

If the laws inherent for reality are absent, how then can we be sure that we know the identity of what we are about to tradeoff? If the identity is absent, then how can we be sure of the worth of the objects I am about to tradeoff? That is, if I have no way to identify the worth of, say, a "beautiful sunset" beyond my own appraisal, then how do I know the worth of my tradeoffs? Perhaps my tradeoffs carry with them a reduction in the quality of life? In fact, I believe that when I exchange the sublime—a beautiful sunset—for money I tradeoff incommensurate goods, that is goods of uneven worth.

The principle of democratic consensus is not explicit enough to call just. The majority of drivers may implicitly want urban America to experience exhaust fumes. This is not justice, however. No supra-economic or technical principles that could structurally aid us in our assessment are apparent. If "the public" wanted mass extinction of other species, as apparently it does (according to its acts), the pragmatism and subjectivism of the realist could not resist it inherently. Indeed, mass extinction does make us literally more "happy" because it provides more utility. Through our expanding styles of living, we enhance our wealth, hence our happiness, or utility (so the neo-classical economist tells us) by cutting down our forests and reaping their harvests. "What profit is it for a man (or woman) to gain the whole world while losing his/her soul?" might be an appropriate question to ask in this regard. Only by appealing to a suprarational or suprasubjective principle could a plea for preservation be made.

Ethics, it must be remembered, has to do with how we ought to live, or more preciously how we ought to harmonize different sides of our lives. Without a structural view of the diversity of life, we cannot be ethical to the degree and to the extent that we must be because ethical principles remain hidden. Principles remain hidden because reality is obscured. If TA wants to help us debate and discuss the particular "goods" and "harms" of technology and also move us from a naive, value-free technical evaluation to ethical or normative deliberation, then it must rigorously unearth the principles that govern reality.

By putting the problem in terms of goods and harms, TA *could*, but too often does not, intrinsically put an ethical or normative question to us. We do not know by what criteria something is considered to be "good" or "bad" if we do not know the community and its principles out of which an evaluation arises. The question of whose particular goods and particular harms are being considered is relevant—if it were asked. People are not merely socially unconnected individualistic units of utility maximization. We arrive and continue as members of groups: families, states, businesses, places of worship, and so on. If a principled public discussion about the sociological place of technology could take place wherein the demands of technology are being weighed over against other of life's demands, then we would all be served.[43]

Sadly, realism is still mesmerized by what I call American majoritarian expertism. Accordingly, only one or a few scientific experts are deemed competent to represent "the majority of the scientific community." Their technical training and experience further intimidates public response. In fact, we come to depend on experts for their opinion and hence abdicate our authority. The consequence of this abdication is an erosion of wisdom that exists prior to expert work. Hence both the citizenry and our government have come to depend increasingly on a regime of experts. This authoritative state of affairs is called expertism.

Axiology—the study of the specific principles that define the ought—could be advanced, furthermore, *if* the "good" and the "bad" were more precisely defined. The rigor that shapes the scientific parts of TA must be matched by an equal degree of ethical or normative rigor and precision if life in its fullest sense is to be enhanced. In reality, economic growth as outlined in the optimism chapter represents to our thinking *the* dominant value that guides assessment. Should economic growth occupy this much place and emphasis? This ought to be the ethical question we ask.

Ellul's critique returns to haunt us. How should we respond to the methodological bias that says that the technical ability of TA can control technique? Is this ability another form of technological domination? The answer to that question depends on one's view of evil or domination. I believe that TA can and has helped us to increase our control over specific *micro* practices of technique.[44] Identification of problems, predictions of outcomes, management of options, and the increased awareness of the principles that undergird our individual technological projects, enable us to strengthen our sense of responsibility. To the extent that we do control technology, we provide a powerful antidote to the pessimists' critique of technology. Technology is not autonomous; we can control it! The control of DDT, and the Concorde Super Transport airplane are but two examples of our ability to control technology.[45]

TA can and has helped strengthen democracy. Disparate parties can adjudicate grievances and thus give increased legal definition to harms and their costs. The Love Canal residents' long sought-after remunerations—small as they may have been—put newer, more powerful legal teeth into the word "harm."

Serious defects in realism's design and methodology seem apparent. As we have said, realism attempts to clarify what is technologically "good" and "bad" on a micro level. An ethical exercise on a micro *and* macro level is present, however, though it is not acknowledged as such.[46] The practice of "hidden values"[47] reduces the effectiveness of deliberations. This is especially a problem for the naive realists. Accordingly, the ethically and ontologically denuded science outlined above was made so because of an inadequate view of ontology and axiology. However, philosophical presuppositions like value-free science *hide* operative meta principles such as progress, scientific-technical control, human autonomy, and utility maximization.

Meta principles of good and harm are operative, even act as ideals, but their integral relationship to the entire social, technological, and scientific adjudication process goes unacknowledged. Principles such as "utility" operate unacknowledged on the micro and the macro level. Consequently, vast sums of money are spent, laws enacted, and lives affected without knowing the ethical *bases* of policy rationale. For example, many of us want the "good" result of material security, comfort and wealth; we want progress; we want the blessings of abundant but cheap energy. What we fail to see, however, is the force with which such principles *direct* society into a macro and a micro acceptance of technical deployment and its consequences. Focusing on economic costs and the technical data has blinded us to more foundational assumptions surrounding the place and force of such ideals as progress.

Progress, the notion that equates social development with economic evolution, has served as an ethical ideal, justifying therefore an easy acceptance of technology by too much of the world. Because we have been able to develop newer technologies before their safety has been insured, we have sped ahead believing them to move us along the road of progress or development. The harnessing of the split atom for energy is a case in point. While pursuing the technical "good" of an abundance of cheap energy on the macro or national level and while only beginning to evaluate the effects of this technology on the micro level, we have neglected to ask ourselves about the inherent goodness of "progress," an influential ideal from the beginning. Consequently, we developed no fully safe way to dispose of nuclear wastes. Modern technique increasingly imposes unintended burdens and related clean-up problems on us. Thus, the pessimists are correct: technique has changed the fabric of our lives.[48] If realism is to be truly realistic, then it must articulate hidden values and foundational assumptions.

Moreover, realists are straining at the micro problematic cell while forgetting about the carcinogenic organism. Traditional regulatory approaches to risk and cost analyses have concentrated upon the maximally exposed "average" individual (the cell) located in the vicinity of the danger area. Studying the absorption of nuclear radiation, to use but one example, from a "leak" begs the question of long-term exposure to radiation that comes with living in this part of the twentieth century (the organism). The global saturation of our planet with technology and its products forms one, if not the most critical, threat to our well-being. Technology does cause problems that have accelerated or "synergistic" effects. The effects of environmental contaminants upon cancer on general population strains is a case in point.

Another example is the way chemical build-up in host organisms is measured. Typically, chemicals are assessed by exposing animals to one chemical for a nominal time. This nominalization of risk is not the situation in today's world. Animals (such as humans) are exposed to a variety of chemicals simultaneously, the interactions of which can hardly be understood. Therefore, a variety of chemicals and their interactions and their build-up creates a synergy or a combined effect that is negatively greater than the sum of its parts. We need a literal worldview perspective about the place and imposition of modern technology to assess these synergistic threats.

Nor is this problem confined to the risks of technology; it is evidenced in the measuring of benefits for a given control measure. Typically, risk assessment cannot weigh the benefits of a given control measure beyond the immediate context of the control situation. For example, if carbon monoxide poses a threat for a given area, risk assessors may recommend a given technique to control CO_2 levels. However, macro beneficial measures, like carpooling, biking, and mass transit are missed.[49] Here management is limited to scientific and technical solutions. Social, economic, or ideological solutions are not attempted because a *technical* micro solution to problems is always preferred over a more varied solution. Modern technique functioning as a social force is unquestioned because realists have no view of the macro social effects of technology.

Technical assessment reinforces the imposition of modern technology upon us because technical solutions are always at hand. Technical solutions, rather than social ones, grow because technology is thought to be *the* solution to our problems. Further, by imposing new techniques of control upon problematic technical situations, realists betray their optimism: technique can correct technique. The imposition of technology grows.

An overly optimistic faith in Reason contributes to additional problems. The traditional paradigm of naive realism has demanded a highly skilled person who follows definite, rigorous, universal procedures of method, logic, and judgment. The process of assessment was supposed to be free from bias, ignorance, incompetence, and ideology. If procedures were strictly followed, objective, universal truth that could be publicly verifiable would follow. The exquisite product of experts would replace superstition, a priori metaphysics, religious bias, and well-meaning but largely ignorant social opinion. Scientific truth could and should become the basis for universal thought and action because it was believed to be publicly verifiable and hence publicly applicable. If consensus failed it was supposed to be caused by irrationalities, ignorance, or prejudice, not by the method of Reason itself. The method was free of human taint, so the belief held.

This view of scientific rationality that undergirds naive realism is abstracted allegedly from all personal, social, and environmental factors. Consequently Reason's function in first analyzing then establishing certainty was thought to be removed from the web of human influences held by "subjectivists." This abstraction and consequent ideologically flawless character of Reason led to an equally great but contradictory dialectic, an equal and opposite form of truth. Science and risk assessment were supposed to control the domain of facts, logic, and empirical observation, while politics served the realm of values, emotions, social influence, and symbolic knowledge. Physical reality was thought to be neutral in this positivistic theory. An essentially valueless, objective picture of physical reality yielded equally neutral, valueless facts. Ethics—personal and social values—was juxtaposed to this essentially neutral process. In effect, a dual world of facts and evaluations was constructed whereby experts were thought to analyze neutrally an equally neutral reality without any of their own humanity interfering. The "public" would then interact with the results of the expert deliberations and form a subjective evaluation. Ethics only entered the discussion as an afterthought to basic scientific research.

Yet problems arose when men of equal standing and goodwill began vehemently to disagree. "Who can we trust?" asked the public. Equally respected experts could not agree and, hence, could not solve the problem. Class, professional, economic, and personal biases were then revealed with the result that the obvious humanity of the scientists eroded the myth of neutrality. Myth and image die hard.

So pretentious a claim created an image of power and respect for science. A self-correcting expert whose methods result in truth and whose deliberations about the goods and harms of the increasingly complex nature of modern technology became a modern social force not to be questioned. Can we, the lay public, match the technical expertise of the risk assessor? If not, why get involved in any technical debate? This view of technical rationality has resulted in a weakened democratic involvement in public debate because it disparages nontechnical thought. Speaking of the alleged ignorance by the citizenry, I concur with Rudy Volti that

> this ignorance can limit the ability of citizens and their representatives to control the course of technological change. Most technologies are built on a base of specialized knowledge: if most people are incapable of comprehending that knowledge, there looms the danger that the direction of technology will be left to a small cadre who are, when public choices depend upon expert information, experts and not the electorate will supply and evaluate this information.[50]

This epistemological chauvinism is nothing less than a trust in Reason and its positivistic methods to create a more certain world. To summarize:

what is significant is that the positivist myth of scientific rationality has encouraged the assumption that really there is no such thing as "organizational risks" from badly adjusted or over-provoked individuals, or from malign elites. This profoundly unsociological perspective is consolidated by the faith that conflict and deviation arise in rational institutions only when someone acts.[51]

Thus, the risk assessment process cloaks unexamined truth claims. This problem is most acute in and because of the "magnificent project of Modernity." Modernity—the revolt of the self-proclaimed autonomous person of the Renaissance and the Enlightenment—unleashes an attempt of far-reaching control over nature and culture. This control motive first arose in chapter 1, when I discussed Reason's attempt to subjugate nature so that human freedom and dignity could be enhanced. It is enhanced in Realism's micro technologies that attempt to remedy discrete situations.

Seeking to break out of all suprarational and heteronomously imposed restraints, modern humanity has set out to dominate nature in the name of human freedom. This massive undertaking was motivated by, among other forces, the technological imperative. While no one commandment defines this imperative, its evolution is nonetheless clear. Philosopher/scientist Francis Bacon sought to dominate nature for two reasons: the harshness of life could be overcome, and the extension of the human empire consequently would yield material benefits. Science and technology were means to that end.

Descartes took the project a step further. Descartes was looking for a trustworthy method of thought that had universal applicability in theoretical and practical situations. His systematic doubt and quantification of reality led him to superimpose a rational-mathematical method of thought upon reality. Reality, thereby, could be reduced to its simplest atomic, indivisible building blocks, then reconstructed through the means of the designer's mind. Reason first disassembled then reconstructured reality. Control was *raison d'état* for this method.[52]

A social agent was needed to carry out these tasks of control and expansion. John Locke provided the rationale for the imperial social individual. Locke was trying to identify an atomistic building block, the socioeconomic efforts of which anchored society. He looked to the individual to provide such a foundation. The autonomous, self-sufficient, unencumbered individual became the universal point of first and final reference for Locke. States, markets, even meaning itself was thought to reside within the individual. Nature became simply a backdrop for the autonomous individual to realize his desires through a contract. Thus, "common order arises out of individual desires through agreement, and this control remains subservient to the individual."[53] Therefore, modernity, through Locke, adds an atomistic social agent to its program for control of nature.

Stephen Cutcliffe summarizes the issue of control:

> Technology in its modern disguise began as the Baconian promise of liberating humanity from the hardness of life and endowing it with the riches of the earth, and it commenced with the Cartesian commitment to achieve liberty and prosperity in a mechanical-scientific way. . . . Then it culminated its project with the Lockean insistence that the sovereign individual and his desires would be the recipient of the fruits of technical control."[54]

This attempt at control is not simply the dream of philosophers. The barriers of nature must be overcome by practical human means. The rationalization of nature became the practical means for overcoming the resistance of nature. Overcoming resistance led to the large-scale redesign and manipulation of nature. "The Cartesian design has been carried out in the methodical construction of encompassing organizations, corporations, bureaucracies, and educational institutions."[55] Gleaming high-rises, concrete superhighways, organizational bureaucracies—these and many more forms of nature rationalization are Modernity's attempt at control and reward.

We have arrived at the profoundest of ironies. Modern technology has caused a culture of risk by attempting to control nature and give freedom to humanity. We have erected a defining technology that purports to give meaning to nature, society, and to the individual. However, we increasingly experience the *macro* harms even while techniques are multiplied on the micro level to cope with the effects of technology. The result is that technicism is enhanced. Technicism means that more technology is used to penetrate and to define more of life, exaggeratedly so.

Risk analysis alone cannot solve the problems created by technology for two reasons. First, because the same fund of methodological analyses, logical procedures, and attempts at control—techniques— are used to promote and then to control technology, realism betrays optimism's easy reliance on technique to save us from technique.[56] *The* issue facing us now is not better micro control. It is the place and the meaning of technology itself. Further, the nature of control, the limits of technology, the meaning of the world around us are among the central concerns we must face if technology is to become safer.[57] Because realists do not address these macro and foundational questions, yet another doubt plagues their methodology.

The knowledgeable reader may object that I am confusing science and its procedures with technology and its procedures. Not so, I would argue. To again quote Cutcliffe: "The same sort of formal logical procedures and methodological analysis are simply transposed from scientific theories and applied to technological innovation then to its assessment."[58] Risk analysis pioneers believed that the same procedures and methods that were applicable for science could be applied to risk assessment. The resultant certainty and trust placed in science was thereby transposed to risk assessment.

Yet another problem for realism is its assumption about what constitutes social and individual utility. Cost-benefit analysis contains a view of the person, happiness, and a morality. A rational, self-interested, utility-maximizing individual seeks to gain utility or advantage and avoid disutility or loss. Utilities are believed to be quantifiable bits of economically and technologically defined happiness that are measured in the market or by law. Realists say that utilities or goods as well as harms can be quantified: the maintenance of the beauty of one sunset equals X dollars. To protect this beautiful sunset, one must be willing to add or to subtract the cost of maintenance of beauty against the disutility of payment. The benefit of beauty is thus measured against the, (usually) monetary cost to maintain aesthetic beauty. Money is measured against beauty.

The rationalistic utilitarian assumptions behind realism need to be questioned. Consumers *always* want more, according to this model, because our self-interest is unlimited. We want more because we are utility maximizers. We maximize our utility because more economic benefits inexorably lead to more happiness, so the theory goes. We are ethically bound to pursue this path because utilitarian ethics, the ethics of utility calculation, says that the only "sovereign" master we must obey is happiness and the only sovereign master we must avoid is pain. This myopic, rigid view of happiness needs to be challenged with a richer, more flexible view of happiness.

Social and personal utility are calculable entities. A realist would argue that we are able to calculate our individual or social utility for two reasons. First, the utility of one unit of aesthetic beauty equals one unit of economic value. This unit usually is measured by a dollar amount. It is believed that we are able to calculate the tradeoff between the disutility of money surrendered versus the utility of a beautiful sunset maintained. This calculation is possible because we are thought to be primarily rational beings who constantly seek our rational best interest. Given our national propensity for environmental degradation and addiction to drugs, the assumption about our ability to calculate our own best interest seems naive if not overly optimistic.

Utility maximization is the driving force behind the ongoing need to control nature and, thus, wrestle from it wealth. The alleged universal push for ever-increasing amounts of economic and technical happiness produces global and intensive expansion and imposition of modern technology equal to or greater than our ability to control technology. The cruel and destructive irony and impossibility of trying to cope with technical problems while at the same time demanding more technology, the foundational assumptions of which remain hidden, *must* not be lost on us. We create a methodology that wants to control the micro harms of technology even while the macro harms are fueled by our acquisitive nature. This fact is especially true for Americans. Thus, there exists an inherent contradiction within realism between expansionary and controlling forces.

It is in this notion of the scientific control of an inherently problematic reality that we locate one significant philosophical root of Realism. This root is pragmatism. The scientific method, a *dogma* according to pragmatist William James, brings control and certainty to the problematic situation. Pragmatist John Dewey openly talks about the "quest for certainty" brought about by the pragmatic method of united thinking—scientific rationality—with doing or applying the methods of scientifc control. Scientific means and methods are used to bring about certainty for an uncertain situation. The status of this method of control is "supreme". The end result of this method of operation is to be, according to Dewey, is *a common faith.*[59]

Is it proper and beneficial to want more technology, even if we could pay for it? Doesn't the inherent pull for a never-ending amount of technological good push us beyond the micro managing abilities of realism? Relatedly, don't we impoverish life by expanding technical reality for two reasons. First, we do violence to nature by trying to control it. Second, we do violence to ourselves when we truncate the meaning of happiness.

There is, as I have just said, an implicit view of happiness being proposed in realism. Utility or happiness comes about by increasing measurable bits of technologically and economically conceived happiness. I do not deny that technology and economics are indispensable for happiness. This is not the question. Rather, can happiness be defined *fundamentally* in economic and technological categories? I am arguing that to do so brings about a reductionism and therefore a narrowing of life possibilities. Would it not be better to attempt a harmonious blending of all aspects and their needs into a more or less holistic understanding of happiness? This question will occupy structuralists, whom I will discuss in the next chapter. For the moment, I may say that in spite of realism's attempt at a broader, more holistic view of preserving, through tradeoffs, the many goods of life, realism's utilitarian *pre*-assumptions foreclose on that more holistic possibility.

The utilitarian maximizing paradigm forces us to view tradeoffs myopically in an overly economic manner. The economic cost is always measured against supra-economic benefits: what are we willing to part with economically to secure a noneconomic end? This question represents the crassest of materialisms because some things do not have a price. This crass materialism suggests incommensurate weighing. Beauty has no price; only pollution removal that can cloud beauty does. This reality should be kept in mind as humanity plunges headlong into the rain forest and thereby destroys hundreds of rare, unclassified species.

Human value is equated with economic costs in this utilitarian ethic. Suppose that one's beloved relative dies a slow death because of complications resulting from proximity to a paper mill that was putting toxic chemicals into the city's drinking water. The legal action is swift and strategic enough to force the company to pay financial remuneration (which is a nearly unwarranted supposition) for the lost loved one. The important question is, exactly how much is your dearest loved one *worth*? The question is entirely warranted because money acts as a common yardstick—a common yardstick for morality—to measure the store of value. Can a price be put on such intangibles? Indeed, doesn't the thought of trading irreplaceable relationships with money cause ethical nausea? I hope so.

Compensations for technological damages often are fixed in terms of "expected life earnings." Therefore, an elderly person will not have the value of a child prodigy because the "expected life earnings" of the youth are much greater than that of the elderly person. Therefore, the courts have awarded larger amounts of money to the young than to the elderly after technological damage because their value or "expected life earnings" is greater. In a society that valued wisdom and age more than money and youth, a different judicial outcome might be expected. Because money is the common store of value, we value earning potential more than age.[60]

The specific problem manifest is a reduction of the profoundness and multidimensionality of life to economic maximization. "Life" manifests a much greater complexity, richness, and fullness than the economic utility paradigm would suggest. The quantifiable, the marketable, the tangible realities of life are but a few modes comprising the rich meaning of life. Perhaps in this context Christ's words, "Surely life is more than food, the body more than clothes" are relevant.[61]

Market prices and court verdicts compensate for loss of analytically distinct bits of reality at best. There is no replacing grand and sublime losses of, say, "beauty." Instead of asking the cost-benefit question, "How much would you pay for a beautiful sunset?" we might better ask what tradeoffs are a variety of social institutions—customers, management, labor—able to make for a reduction in the amount of pollution or harm at the margin. That is, given scarce resources and a diminishing marginal return on our pollution-reducing dollars, what other preferable activities are specific relevant communities collectively and individually willing to forego not for beauty (for this has nearly infinite worth) but for less pollution that obscures this beauty? Might we be willing to pay one hundred dollars more for a car, say, if that extra money paid for a reduction in auto emissions?[62]

Beauty and life, like other intangibles, have profound and nearly infinite worth and importance. This worth must not be confused or equated with economic value. Economic compensation cannot possibly measure the loss of any amount of life because life is of nearly infinite value. The profundity of life should cause us pause when we contemplate the future of our technologies.

I continue with the question, "How much would you pay to attain some good or avoid some harm?" Framing the question in this manner creates a "shadow market." A shadow market is one in which in some imaginary future time one will commit funds for projects one believes are worthy. For the moment, no real commitment is required. Since one is not literally paying out current scarce resources, only future or shadow resources, one's desire for purity often is greater in the present when one does not have to pay than in the imaginary future when we will have to pay. When we don't have to pay for technological projects, but are only contemplating them, we optimistically may believe that we can maintain them safely and pay for their upkeep and waste disposal. In this case talk is, literally, cheaper than commitment.[63]

Sometimes tradeoffs involve difficult choices. For example, the state of Washington had to choose ten years ago between employment for loggers and the beauty and integrity of virgin timber. If residents kept the timber then the loggers would be without employment. If they allowed the loggers free reign then the natural beauty and attendant limited wildlife would perish. The state was at an impasse until a suggestion came forward the effect of which was to transcend the difficult situation in favor of a normative or ethical condition that enhanced the quality of life for all. An agreed upon stand of virgin timber was off limits for cutting. Other stands could be cut. Consequently some loggers had to be laid off while many remained. Unemployed loggers were offered other jobs serving as rangers who would plant trees and thus reforest timber stands. Money to employee these people were taken from a one penny increase in gasoline tax and a user fee to enjoy these virgin stands of timber. Thus, the principles of conservation, employment, and justice for the land and loggers helped transcend the difficult situation.

This shortsightedness is most visible in the rise of the nuclear industry. We were led to believe the industry, including the cleanup, was safe, and therefore committed vast amounts of resources to it even before we had a sufficient picture of the true long-range costs. In this case, the "shadow market" is the unanticipated costs entailed in the clean-up. We confidently initiated the nuclear power industry and thought we could maintain it because at the time we did not have to pay literal clean-up costs. Now that costs for clean-up are rising astronomically, we realize how punishing the industry has been. The scientific and political sponsoring forces that legitimated fission to the public should now be seen as status seeking and self-interested.[64]

I have argued that science and its offspring risk assessment have been motivated by the quest for knowledge, control, and power. These principles are energized, in turn, by the foundational quest for certainty.[65] Risk reduction is based on the assumption that technologies carry with it uncertainties. The outcome, implementation, and the management of technology carry with it uncertainty.

It has been my argument that the origin of certainty, not the reality of it, represents a decisive problem for risk assessment. Whether this origin is located in a subjectivistic, democratic populace or a rationalistic, objectivistic expert, a faith claim nonetheless is involved. There is no empirical, unbiased test that demonstrates with absolute clarity whether "the people" or experts provide the surest, safest origin of principles or theories that can manage technology. Dogmatic claims in the trustworthiness of Reason in the objectivist and of the surety of majority consensus are trusted in as bedrock authority. Therefore we may conclude that faith or certainty exists. That is, faith or foundational certainty, sui generis, contains its own internal set of human demands and self-authentication. Faith sets the conditions for subsequent analysis by forming a foundation of trust. Trust, in turn, secures institutional support through the construction of paradigms. Faith, like Descartes's confidence in rationality, is not subject to rational justification; certainty justifies rational procedures and, hence, leads to acceptance. Acceptance of given frames of reference and paradigms reinforces the acceptance of paradigms; the internal logic of acceptance is circular. Faith thus directs Reason while doing empirical interpretation. Thus, faith or certainty, a universal human function, represents the suprarational expression of certainty present in all theory and life.[66] Indeed, the nature of risk forces us to need certainty as a bromide for uncertainty.

If faith is present, then two consequences follow. First, the ultimate neutrality of science and its cousin, risk assessment, is undercut by this need for supra-empirical certainty. Second, we are not far from admitting another source of supra-empirical faith: divine revelation as a foundation for doing risk assessment. Why should the dogmas of Reason be any more credible than the dogmas of revelation? If divine revelation can in any way be said to offer insight, then a different ethic could assume a crucial role in this entire project! (This will indeed be the case when I turn to structuralists.)

I have argued that a degree of uncertainty or risk is inherent in the project of mega-technology and science itself, both being a product of modernity. This element of risk or uncertainty prods us to need certainty. Yet realism is plagued by its assumptions. The systematic breakdown and attempted reconstruction and control of reality makes the project fundamentally uncertain from the beginning.[67] If we believe that technical experts or scientists can correct this fundamental error, we underestimate the consequences of control. The world cannot become a reassembled machine.[68] If we fail to locate the reason for uncertainty in our attempt to reconstruct reality, then our risk assessment will be fundamentally flawed.

Not only is there macro uncertainty inherent in the project of risk analysis, but uncertainty often attends our humanness: we are limited beings, not able to assimilate all necessary data. "Uncertainty is inherent in risk assessment. This is due to the large number of assumptions and inferences which are needed to compensate for the lack of data. . . ."[69] New and more information often does not make for more certainty. Sufficiently pervasive uncertainty too often undercuts a risk assessment because knowledge of all necessary factors is not possible. New improvements in testing may provide more information. They even may lead one to different conclusions, but just as often certainty is not increased whatever the amount of information available.[70]

To be sure, there is a needed role for systematic observing, calculating, and testing in risk analysis. Thus, instrumental rationality is mandatory. However, this more humble role is not the one claimed by rationalism, the exaggeration of human Reason. Objective, value-free Reason that seeks systematically to remake reality so that its perfectible objects can be realized in an improved reconstituted reality is arrogant and hence oversteps technology's more realizable task. Errors in technology are not the result of random deviations from an essentially perfectible technical object, and thus are correctable by better rational technique. Rather, many technological errors result from either the over-extension of technical rationality itself or the sheer ponderous and unrealistic nature of a given technical project.[71]

> Modern men and women are not content with enough technological control of the world to satisfy their basic and vital needs. They seem to want to penetrate the world itself with a new pragmatic spirit and a sense of their domination of it, so as to experience and represent themselves as lords and masters.[72]

At best, risk analysis can only aid us in estimating risk posed by new technologies, then propose possible alternatives. It cannot give us certainty. Risk analysis can decrease definite anxieties through targeting specific agreed-upon problems that are located often in the biological, chemical, and mathematical areas of life; but certainty is not a matter for science to decide. This certainty—faith by definition—belongs only to that which is believed to be ultimate. Its fixed character defies the relative and limited grasp of *any* scientific endeavor—including theology. Faith's character is more serendipitous than technical, more the product of grace and less the product of rigorous effort. To be sure, rigorous thought can be and often is a *means* to certainty. If certainty is attained in that manner, faith is less the product and more the foundation of theoretical thought.[73]

4

Technological Structuralist

INTRODUCTION

Finally, I turn to the thinkers I call the technological structuralists. This fourth view is different fundamentally from previous positions in one central and three related ways. Structuralists have a different view of the matrix or the context of life. This fundamental disparity suggests at least three additional differences. Structuralists believe, contrary to optimists, that many aspects make up life. Hence, the study of technology should take its place alongside other disciplines when addressing problems. No mere technical solution is permitted. The enduring quality of these aspects keeps evil from being as all-encompassing as Jacques Ellul suggests. Further, they have a more diverse, richer view of the person than do realists. The utility maximizer is much too shallow a view for the structuralist.

Furthermore, structuralists differ from optimists in that they are less confident of technical expertise and its ability to deliver us from trenchant problems. Human shortcomings and evil have impacted technology because it arises out of the human condition. Structuralists would argue that there is no autonomous and neutral technology and thus differ with the optimist. Technology cannot be simultaneously neutral and *the* source of healing for all major human problems. Further, modern technology overburdens humans because we have repeatedly turned to it for deliverance. Burgeoning technology has obscured a more holistic worldview. Cultural saturation with technology is a condition to be avoided. Structuralists attempt to outline and then implement a technology that complements rather than dominates life's many needs.

Structuralists fundamentally do not agree with pessimists. While recognizing that technology does present an overbearing force in modern life, and thus can be totalitarian, structuralists do not see technology as inherently evil or the origin of the loss of modern human freedom. A certain kind of technology has and is making a positive contribution to human life. Further, technology can add quality to life. This idea of the inherent goodness of technology presents the first key to understanding structuralists and is by no means in conflict with their notion of the corruption often manifest in technological projects. Most importantly, technology does not cause social determinism; evil is not sovereign. The structuralist believes that humanity can exert control over technical systems but only by a deep inner renewal of the human condition.

Finally, the structuralist is not a realist. While appreciating the multifaceted way technology is evaluated in realism, the structuralist does not share the realist's utilitarian understanding of human nature. Human beings are not fundamentally rational, utility maximizing, acquisitive, consumers whose basic purpose in life is to maximize "happiness" by calculating the economic costs or tradeoffs of technological deployment. This limited view of the person does not help us arrive at *the* central structuralist question: what place must technology occupy in our lives?

The term *structuralist* comes from *structure*, a universal pattern, law, or standard. A structure gives order, limit, or purpose, while at the same time providing identity to any activity or thing. The structure for a business with its attendant law of careful administration of scarce resources is one example of a structure and an attendant law. A business is different from a family because its identity and character are different. Its limits and constraints will not allow it to dominate life but will ensure its continued existence, as the history of communism attests.

The concept of structure parallels, but is not exactly equal to, the theologian's use of "general revelation," the philosopher's use of natural law, and the natural scientist's designation of positive law. Its results are universal and cannot be denied with impunity.

The structuralists I will discuss are both theists, a fact that will color their view of structures. The following quote by the eminent structuralist Egbert Schuurman of the Netherlands defines one of the central assumptions of the structuralist: "God is the Origin of all things. . . . He binds the creation to His laws. . . . [Consequently,] history manifests a coherence of diverse aspects which are led, controlled, and consummated by God."[1] These "diverse aspects," or "rooms" as I am calling them, are from the simple to the more complex: numerical, spatial, kinematic, physical, biotic, psychic, analytical, technical, linguistic, social, economic, aesthetic, juridical, *pistical* or faith-based.[2] These rooms represent the structures for life. These rooms or structures are not "whats" but "hows"; they are ways of behaving. These aspects have an identifiable, positive, but limiting role to play in reality. Together they represent the study of being (or *ontology*) and form an integral matrix for life.

Since the technical room was created by God as part of the "very good" creation, something of that goodness remains, in spite of our evil and exaggerations. The technical room was created to cohere with the rest of the rooms or aspects of life. Schuurman agrees with Ellul that technology is exaggerated in modern life. However, the limitations God intended for each room are still present, our pretensions to autonomy and the consequent reductionism not withstanding. These structures also give positive identity, meaning, and purpose to the rooms or aspects of life. Structure simultaneously limits yet gives purpose.

The second structuralist, Ernst Fritz Schumacher, agrees with the fundamental importance of structure for technology specifically and life more generally. "Everything in this world has to have a structure, otherwise it is chaos. . . . [Everything] has no meaning without structure."[3]

Structure forms the order and the meaning not only for persons and institutions but for all reality as well. Engineers presuppose it.[4]

Schumacher has a slightly less imposing view of the structures for reality, but the intent is nevertheless the same. Reality is separated into four levels of being with four concomitant levels of knowing, according to Schumacher: "the universe is a great hierarchical structure of four markedly different levels of Being."[5] These levels of being are from simple to more complex: mineral, plant, animal, and man. Accordingly, while Schumacher and Schuurman differ somewhat about what specifically constitutes basic strata in reality, they do not disagree that reality is differentiated because of basic structures holding for reality. Thus, the fundamental concern for the structuralist is not epistemology but is ontology. It is not how we intellectually *know*, but how we *live* and the conditions for that existence that concern the structuralist.

Structuralists talk about placing technology within a context of reality because it is reality that gives meaning for life. Life cannot be constructed; it is given with meaning. Technology must fit into and complement the other aspects of life if life is to be meaningful. When it does fit in and does not dominate, technology enhances life. The main problem today with modern technology is that under the guise of the pretension of autonomy, technology dominates too much of life. Totalitarian technology cannot dominate life totally because life continues to manifest structural diversity, as the liberal arts curricula amply attest. To dominate reality totally would mean a destruction of life and culture. Only so much of the teacher's classroom, for example, can be taken up with technology. Eventually, totalitarian technology will be resisted by an ontologically complex reality.

ERNST FRITZ SCHUMACHER

The academic credentials of theorist/technologist E. F. Schumacher are impeccable. He was a Rhodes Scholar in economics and became an economic advisor for British control of postwar Germany shortly after World War II. He was one of the top economists and head of planning for the nationalized British Coal Board of England for over twenty years. His company's abuses in mining led him to insights that caused him to raise his voice of concern for soil conservation. This concern, in turn, led to the presidency of the Soil Association, one of Britain's oldest farming associations. It was during his work with soil conservation that he was forced to deal with the deep effects of modern technology. These effects troubled him enough to became founder and chairman of the Intermediate Technological Development Group, a think tank for alternative technology. Returning to his first love of economics toward the end of his life, he became director of Scott Bader Company in England. This business emphasized worker ownership.

The fundamental problem of modern technology is "the idolatry of giantism," according to Schumacher. "Today we suffer from an almost universal idolatry of giantism in technological development."[6] Modern technology is too big because it usurps the legitimate place of other forms of life or different forms of human work. It seems natural for a structuralist to locate the fundamental problem of modern technology in its *place* or in its intrusion upon other necessary areas of life.

Giantism is caused by our desire to control and coordinate reality, says Schumacher. Its primary symbol is the Industrial Revolution. The Industrial Revolution is the mass production and consumption of a large number of goods. It is centralized, capital-intensive production on a large scale. Such large-scale production encourages the exploitation of nature. Mass production is also energy intensive and labor displacing, and it tends to concentrate power and wealth in too few hands. It promotes meaningless labor, meaningless lives, and a loss of faith in ourselves and in others. This fact is sadly ironic since reality is created to manifest meaning and purpose. Thus, meaningless technology distorts human experience. However, Schumacher is not an overwhelmed pessimist.

> I have no doubt that it is possible to give a new direction to technological development, a direction that shall lead it back to the real needs of man . . . to the actual size of man. Man is small, and, therefore small is beautiful. To go for giantism is to go for self-destruction.[7]

This quote demonstrates two crucial points in Schumacher's position. First, unlike the pessimist, he sees a real possibility for change in the direction of modern technology. Indeed, his disciples have erected alternative technologies around the globe. Second, because humans are believed to be small, technology must be limited to fit their size. In a moment we will see that there are structures *outside* of humanity that make this limitation possible. Deny these structures and we commit self-destruction.

Modern technology narrows our cultural choices. We measure the kind of world we live in—First, Second, or Third—by degree of industrial development and progress. It functions as a standard in the sense that a nation seeks to emulate the best—that is, First World—standards. However, one key consequence of emulating First World technologies is an erosion of Third World cultural patterns, as Schumacher argues: "First World development leads to breakneck urbanization, heavy capital investments, mass production, centralized development, and advanced technology . . . all of which results in impoverishment for most Third World residents."[8]

Technological development is not the only source of the problem. Traditional economics, or "imperial economics," contributes to our woes as well. Economics is imperial for two reasons. First, to behave "uneconomically" is to incur moral approbation of the first order, says Schumacher. This rigid yardstick is currently being applied in higher education. If one's department does not average at least X students per class, budgets are cut and faculty are reduced because you are inefficient.[9] Second, "the sole criterion to determine the relative importance of goods is the rate of profit that can be obtained by providing them."[10] These economic mandates promote the overextension of technology because technology is used traditionally to increase economic advantage.

This economically dominated mind-set has its consequences for nature as well. Nature is defined as raw material, the official good of which can be located in its service to the needs of industry. Nature is treated violently because it must be conquered before its wealth is exploited. Technology represents the violent means by which we secure the economic rewards we want from nature. Consumption leads to "happiness" or utility and is a reward for the disutility of labor. Consumption degenerates into obsessive compulsion that is based on an economy of habit-forming consumption—so argues Schumacher.

This marriage of economics to technology has become so firm that the pair functions as a way of life, not unlike medieval religion. Again, Schumacher says: "What is quite clear is that a way of life that bases itself on materialism, therefore on permanent, limitless expansionism in a finite environment, cannot last long."[11] Modern economics and technology have become wedded to the economic materialism of our age. I consider that fact quite ironic for a capitalistic system, because it was Karl Marx, the "father of scientific [atheistic] socialism," who is generally believed to have developed the philosophy of economic materialism. Schumacher says of capitalism that it too will fail just like Marxism because of its mindless commitment to economic materialism. Schumacher establishes his analysis and critique through the deployment of philosophy.

Schumacher was not a trained philosopher. However, he was deeply reflective and gifted enough to construct a life philosophy that framed his thoughts about modern technology. His philosophy was culled from practical and theoretical sources. His field experience in setting up alternative technologies, his managerial and corporate experience, and considerable theoretical acumen led him to reflect systematically upon the foundations of life.

It is in *A Guide for the Perplexed*[12] that he develops his life philosophy. The book argues that there is a structured, layered reality that undergirds all of life. This structure, called natural law, gives pattern and order to reality. This structure or pattern provides a map or a guide for life that can be followed and thus gives us direction, not perplexity.

There are four levels of being and four concomitant "great truths," according to Schumacher. The four levels of being are mineral, vegetable, animal, human. The great truths will be outlined in a moment. Structure makes possible these basic levels of being.

Further, the complexity of life increases from the simplest to the more complex levels of being. The "lower" kinds of life and being are not less important than higher being—they are just less complex. Lower life is included in the higher modes, but the meaning of higher life cannot be reduced to that of the lower modes. For example, humanity is composed of molecules but we cannot reduce the complexity of our humanness to molecules in motion. Schumacher, like Schuurman, is against philosophical reductionism or the reducing of structural complexity of life to its lowest elements. This graduated complexity represents a hierarchy or levels of being and is the first great truth for life. God is the Author both of technology and this complexity that forms the context for modern technology.[13]

The second great truth is that man is "adequate" to apprehend or understand knowledge of the world in which he lives. Understanding can result in a responsible ethic. A meta-ethic or macro-ethical worldview can result from our technological and scientific knowledge. Ethical obligations clamor, in the evidence, to be heard.

The third great truth stems from the first two. There are four broad fields for study that result in scientific facts and concomitant ethical responsibilities. They are the inner world of myself, the inner world of you or the other person, the outer world of myself, and the outer world of the other person. Proper human development includes a more or less even awareness of these four fields. Schumacher argues that myopic education, such as traditional engineering education, that focuses on the outer world to the detriment of the inner world is problematic.

Different kinds of problems require different kinds of solutions. This is the final great truth gleaned from a study of structured reality. One kind of solution cannot possibly cover all kinds of problems because all reality is not of one kind. Thus, hunger cannot be solved merely by technology in the same way, say, that an automobile's faulty engine can be technically corrected. Hunger is a much more structurally complex problem. To attack hunger is to marshall many levels of being to the problem. Believing that the hunger problem has a mere, or even primarily, technical solution is to commit a grand reductionism or a simple-mindedness, according to Schumacher.

His concern for the levels of being or structured reality strengthens his critique against the "giantism" of modern technology. Modern technology has overstepped its boundaries in its imperialistic lust to dominate all of reality. Humanity has attempted to make nature, indeed all of reality, subservient to our autonomous agenda by the means of instrumental reason.[14] This instrumentalization of nature has resulted in the desecration of nature and the retardation of the human character. Instrumental reason, or reason seen only as a technical means to some useful end, has become the end for life.

A richer, deeper view of reality and the person is needed. Schumacher would certainly affirm the scientific need to understand the world through the use of our minds and our senses. However, we do ourselves an injustice if we reduce knowledge to mere outer physical meanings.

> Our bodily senses . . . do not put us into touch with higher grades of significance and the higher Levels of Being [sic] existing in the world around us: they are not adequate for such a purpose, having been designed solely for registering only the *outer* differences between various existing things and not their *inner* meanings.[15]

Thus, understanding the physical facts represented by the first and outer level of understanding is but a first and a superficial step. If this were all the knowledge gained, one's understanding would be inadequate, argues Schumacher.

Many times in technological studies we reduce ourselves to this initial level of knowledge. When we think of reducing hunger in the Third World, we think of crop yields, capital flows, calorie rates, and output. This knowledge is all necessary but only preliminary. The next level of knowledge, one needing but not reduced to the first level of knowledge, is a movement to the meaning and the significance of life. It is knowledge of the object as a whole, or knowledge that is informed by the four levels of meaning listed above.

This knowledge concerns itself with understanding the interrelatedness of all levels of being. Humans may be the most complex kind of being, but they need and incorporate less complex levels of being. Without water plants, minerals, and humans could not survive. We harm or destroy ourselves when we harm and destroy the natural environment because we depend on the environment for our lives. The harm we bring to this level of reality affects similar realities in us because part of our nature incorporates natural dimensions of life. Harm the trees and we harm our need for oxygen. This harming is part of the "violent" manner of modern technology; violent because we do not know the sublime principles that must accompany our use of technology. Love, justice, self-awareness, wisdom, and gentleness must become as sure a basis for knowledge as is first-level sense knowledge of outer physical characteristics.[16]

Schumacher claims that the goals of the first two levels of the inner world include successfully realizing the principles wisdom, harmony, ethical responsibility, faith, and service. While knowledge of the third and fourth levels—empirically based knowledge of the outer world— leads to control or mastery of the physical world, as well as to a prediction and a manipulation of human character, truly adequate education incorporates *both* kinds of knowledge in a way that is complementary. Faith and scientific reason must be friends; sight and intuition must harmonize, not battle, he argues. When these different forms of knowledge *complement* rather than compete with each other, technology will complement or fit into life rather than dominate it.[17]

Complementary knowledge leads to a different kind of technology. Rather than man serving machine, as we have essentially done since the Industrial Revolution, the machine will serve man. Mahatma Gandhi's small hand loom for spinning cotton exemplifies this kind of technology, according to Schumacher: small, labor-intensive, gentle, low-cost, consumer-friendly.

By small he means that it fits the size of humanity's relative needs. It is judged adequate by an inner wisdom that enhances but does not dominate other outer needs of life, such as a clean environment. It is labor-intensive because our outer-world need to work, while good, is not set over against our inner-world need to be gentle. This kind of work is certainly not a "disutility" or an evil to be technologically removed. It is gentle because a complementary view of nature will suggest that technology must enhance not destroy the environment. It is not capital intensive and hence does not cost much. It is consumer-friendly because it is designed to adapt to human needs.

Only a whole person can be adequate to the task of apprehending complementary or holistic knowledge.[18] The higher the level of being, the higher must be one's awareness of both inner and outer states of being. Behind much of today's "mindless consumption" is a dulled consciousness or awareness. We are not aware of ourselves, argues Schumacher, because the work we do has become boring and deadening—and this is because of the mode of technology employed in the workplace. Automated workplaces kill human self-awareness, argues Schumacher throughout his works.

An inner and outer awareness for all of its value does not give us adequate knowledge. The outer world also gives us a social world. Schumacher argues that the chief virtue for the social world is altruism. Altruism is based upon the sentiment that one has "enough" resources, is adequately supplied, and consequently one has enough to share with those with whom we are linked by the nature of our collective social humanity. There exists a social awareness or a feeling of empathy because a common bond results from our common social nature. If empathy dies or is retarded, then technology becomes brutish. The Industrial Revolution and child labor, argues Schumacher, is the primary example of this fact. Technology need not be this way, however. Technology can be designed to meet the human needs of the user. Altruism takes place when an engineer considers the needs of the user *first* before other factors. I believe that modern ergonomics[19] is a product of altruism.

Study of the history of technology reveals that technological knowledge is often thought to be pragmatic, workable, instrumental, and utilitarian. Schumacher would not deny this kind of knowledge; he would simply add to it. Schumacher argues that faith is determinative also for all of knowledge. This is especially true for a comprehensive theory of knowledge. He postulates that comprehensive systems of knowledge contain several key hidden assumptions accepted without empirical verification. There is both an exaggerated truth claim (species adapt, therefore *all* of life is subject to evolution) and a hidden untested rule of belief (seeing is the *only* way to believe). Thus, "faith" uncritically holds on to an untested basic assumption: "descriptive science becomes unscientific and illegitimate when it indulges in comprehensive explanatory theories which can be neither verified nor disproved by experiment. Such theories are not "science" but are 'faith.'"[20] Perhaps the grand assumption of science is that reason is capable of sustaining more than limited truth.

I raise the issue of faith and science because I believe that Schumacher is touching on a reality that is apparent in technology studies and was manifest in optimism. Optimism cannot conclusively prove by empirical means that technology has significantly improved such chronic human problems as hunger without causing equally great problems. However, because they are certain, and therefore have faith in technology and technical Reason, optimists believe that technology can be trusted to end lasting problems. The unfounded certainty is projected onto the tool and the results are believed to be similar.[21]

While Schumacher would not deny that a level of pure objective, physical knowledge is possible, this level of knowledge cannot be separated from the more noble world of meaning and purpose. For example, it is not enough to be content with having two cars, thus fulfilling part of the American Dream. We must ask ourselves: what meaning and purpose do those cars hold for us? Are they status objects? Does our enjoyment come at the expense of harried work? Are they "chariots of individual freedom," the enjoyment of which tears at community? Or are they means of service? One's views of meaning and purpose, says Schumacher, will inform automobile choices.

Schumacher concludes that there is one highly significant field of knowledge. He argues that scientific investigation presents us with two types of problems: convergent and divergent. Convergent problems force us to look for solutions to problems that may seem contradictory but are in fact related. A holistic worldview seeks harmonious relationships between technology and the rest of life because technology can converge with life to complement the rest of life. Indeed, the structuralist generally, and Schumacher especially, may be characterized by his or her attempt to contextualize modern technology. Humans need complementarity because we need unity within the world. Schumacher is not advocating a false unity based on a technologically controlled society. Reality does not have to be arranged to be whole; reality is whole and we must cooperate with this wholeness.

Similarly reality often manifests life-enhancing divergence: growth versus decay; necessity (natural law) versus freedom (growth). The same dialectic or opposing forces are at work in the thought of Jacques Ellul. Schumacher argues, however, that this tension seeks a holistic solution to the problem at hand. Leaves grow, then decay. In their death en masse there is life-giving mulch. Technology can merge opposites to create healthful unities. A mulching lawnmower blade is an example of this point.

What we cannot do, argues Schumacher, is attempt technically to manipulate reality so that all divergences disappear. Schumacher here addresses the optimist. Freedom cannot be found primarily in the technical manipulation of nature. Rather, freedom is something that is fixed and a given for life. It is part of our "higher," more humanistic nature, while our body is part of our "lower" nature. When we speak of freedom as the technical by-product of the manipulation of reality, viewed as calculable, manipulable bits of discrete matter, we do violence to ourselves and to nature. This view of freedom leads to a reductionism.[22] Thus,

> While the logical mind abhors divergent problems and tries to run away from them, the higher faculties of man accept the challenges of life as they are offered . . . knowing that when things are most contradictory, absurd, difficult, and frustrating, then, *just then*, life really makes sense.[23]

Reductionism engenders a loss of freedom: "Our ordinary mind always tries to persuade us that we are nothing but acorns and that our greatest happiness will be to become bigger, fatter, shinier acorns; but this is of interest only to pigs."[24] Schumacher is arguing that the scientific mentality that undergirds the practice of technology has created a view of the person and a rationale for technological use that is myopic. In this he repeats the critique of optimism made in the first chapter.

Because humans are defined accordingly as material, biological beings, our most important needs are thought to be material and biological, included in which is energy. When natural fuels seemed insufficient to meet the nearly insatiable needs necessitated by this materialistic reductionism, we turned to nuclear fission to supply our power needs. All was thought to be fine until the problems surrounding safe waste disposal of spent materials proved daunting at best. Thus, our technical knowledge was sufficient to split the atom while our "higher" more humanistic understanding of care for the environment, and a just distribution of pollutants, has escaped us to date. Schumacher argues that life consists in more than the manipulation of matter.

PRACTICAL ALTERNATIVES

Schumacher's theoretical and practical work in analyzing technology's impact has spawned a dedicated following of competent practitioners intent on implementing his ideas. Nowhere is this fact more apparent than in the work of George McRobie, a friend and colleague of the late Schumacher's.[25] The strength of McRobie's work, especially for the beginning student, is the thorough listing and sympathetic critique of Schumacher's basic principles. Technology must:

- be in harmony with the environment, not violent
- use renewable resources
- be simple and inexpensive to maintain
- not be capital-intensive; not be large (a term McRobie will explain), yet *not* be primitive
- be labor-intensive
- be easily sustainable
- be geographically diversified, not concentrated

- use local materials
- be constructed by and made for local workers
- be holistic in its fit with the rest of life
- use natural energy sources such as wind, sun, and water
- be utilized especially by the Third World
- be developed creatively and not produce boredom in work.[26]

One of McRobie's chief roles is to field criticism of Schumacher's work, clarify when necessary, and expand when needed. This he does well in passages when he shows how a different kind of technology will spawn a different kind of economic and political ethic.

The vision for a small technology necessitates a newer, more decentralized, political order. McRobie argues that a centralized social order is the product of a centralized technology such as that of the Industrial Revolution. Centralized political, economic, or technological power is antithetical to the small-is-beautiful outlook. Centralized technology and resultant life marginalizes too many people and thus robs them of the responsibility of caring for themselves. Thus, "orthodox, conventional economics, technology, applied science, and administration are designed to serve an efficient system of production and consumption and not to develop the capability of people to look after themselves."[27] McRobie argues that alternative technology will do much to promote human well-being. Indigenous technology will enable the Third World, for example, to lessen its technological dependence on costly First World technologies, thus promoting socioeconomic independence. Fewer capital-intensive technologies will also preserve the environment and save wear and tear on the local infrastructure because the *size* and thus the *stress* is more moderate—a fact unknown to most modern Americans who use the roads.

"Labor-intensive" technology draws from a local pool of labor for work. Thus, problems associated with unemployment and idleness are addressed. The concept of a labor-intensive technology is particularly applicable in contemporary urban America. There are alternative projects, ones initiated by Schumacher's disciples, dotting the ecotechnical[28] landscape in America. Thus, there is hope of changing the influence of technology. The general trend to self-sufficiency, hand-built projects, even increasing amounts of home-grown food testify to this change. This trend is prompted by the need to humanize technology with the needs of humanity and nature.

Schumacher's insistence on home energy conservation in the mid-1960s when cheaper fuel predominated made him a progressive for his time. Included in this conservation move was home heating through solar energy and solar greenhouses for the year-round growing of vegetables. These practices are accepted more now. (I expect to see solar energy growing in popularity.)

McRobie argues that many in America are moving to a smaller-scale, lower-cost, self-help kind of technology. Community development associations and cooperative movements tend to encourage this kind of technology. A small-scale, locally owned business that gives greater power and responsibility to the worker is also a positive step because it mitigates the need for larger technology. Similarly, self-sufficient smaller cities are dotting the landscape. These cities emphasize cooperative buying, public transportation, public or common grounds, as well as food and water self-sufficiency. McRobie argues that these kinds of projects provide a necessary alternative to the increasingly concentrated forms of technology experienced in typical American cities.

Communities throughout North America have begun to share information on newer ways to develop alternative technology.

[T]hese networks represent many facets of growing revolt against giantism in today's economic and political life, a way of life that increasingly begets violence and consumerism. Recognizing that any real change in contemporary industrial society must start with the individual, with changes in personal beliefs and lifestyles, its members are trying to live now as they envisage the future should be.[29]

Thus, Schumacher's views of technology seem to be influencing American life.

Schumacher has invested most of his time in developing alternative technologies for the Third World, a fact that McRobie amply demonstrates. Schumacher was especially adamant about helping to create alternative technologies in the rural areas. Typical urban technology represented all that was wrong with modern technology: capital not labor-intensive; too big and hence too violent; ecologically damaging; geographically concentrated; dependant on nonrenewable resources; unmindful of the needs of the worker; and insensitive to the local living conditions. Indeed urban technology, rather than creating democratic opportunity, creates opportunity for a few and slums for many.[30]

Schumacher implemented his first alternative Third World technology project in Kenya in 1963. The successful Kenyan experiment led to the development of a low-cost, labor-intensive sugar plant in

India. His model factories consequently were requested throughout much of Africa. Schumacher found out that one large sugar plant required the same capital investment as did forty mini–sugar plants. Further, as he monitored the production of the forty plants, he found that these mini-plants produced two-and-one-half times more sugar than did one large conventional plant. Indeed, the smaller plants employed ten times more people than did the larger mill. This experiment enhanced employment and the sociocultural life by simultaneously building employment and maintaining traditional mores.[31]

Word of this miracle spread to nearby Pakistan. By the mid-1970s, Pakistan witnessed a dramatic multiplication of alternative projects. In the areas of food and agriculture, energy, waste utilization, agricultural implements, and energy generation, the concept of alternative technology was to change field cultivation.[32]

Ghana, in West Africa, represents perhaps the most dramatic rise of local technological self-sufficiency. This nation made glues for export out of local materials. They used local labor, of course, and even used local venture capital for start-up costs. Resulting employment, circulating capital, and reduced production costs created a permanent, feasible technological alternative for Ghana.[33] Schumacher, again, received credit for this work.

In Nigeria, making better labor-intensive tools seemed to be the primary need. Farmers needed tools such as hoes, hand-generated threshing machines, and farm carts. The medical community needed, in turn, self-propelled wheelchairs, bicycle-drawn ambulances, and crutches. Schumacher helped create factories that produced these needs. Alternative technology spawned teamwork, socioeconomic independence, more equality, and hope.

Schumacher's alternative vision also spawned newer ways to view work and labor. To understand what good work is, one must first understand what Schumacher means by bad work. Bad work starts with bad technology. Traditional technology leads to boredom, unemployment, and heavy debt because of its repetitive, labor-displacing, capital-intensive nature. The compensation for the boredom or the disutility of work is the utility of consumption fueled by expanding wages. A consumption-oriented society intensifies the realities of ecological violence, depletion, and unemployment—all of which is especially inappropriate in the Third World. Traditional technology undermines the integrity of work and of life. The basic aim of modern industrialism is not to make work satisfying but to raise productivity;

its proudest achievement is labor-saving, whereby labor is stamped with the mark of undesirability.[34]

Further, capitalism believes that the primary motivation for work is self-interest, which Schumacher argues too often degenerates into greed or selfishness. This lust for pecuniary gain leads to organizational authoritarianism because the worker has allowed management the exclusive right to determine the workplace. The worker thus is robbed of a sense of responsibility and democratic input. This imperial management-centered system can only hold out an economic incentive for work. A more dignified reason to work might be self-realization or service to humanity. This dignity is too often absent today, according to Schumacher. Capitalism could only mock, therefore, a belief that argues that work could have a spiritual aspect to it—an aspect that aids our creativity and enhances broader vistas for our lives. He calls this economic myopia an "idolatry" or a worshiping of economic production and consumption.

> The degeneration of the industrial system—that is, its ever-intensified idolatry of getting rich quickly—offers everywhere ample opportunities for bringing light into dark places. Everywhere the values of freedom, responsibility, and human dignity have to be openly affirmed, even while the neglect of these values would appear to allow the big industrial machine to run more smoothly and efficiently.[35]

Good work is defined by these enumerated principles that form the foundation for life.

Schumacher is not a socialist. Socialism would lead to bureaucratization or organizational centralization. He has fostered in the small and large businesses he has directed a democratic governance structure. That is, management seeks opportunities to consult with labor on the contours of anticipated projects. This democratic view redefines the role of bureaucracy within the organization. Schumacher argues that the principle of subsidiarity best governs any bureaucracy. This principle states that authority must be passed to the lowest practical level so as to minimize bureaucracy and maximize a sense of worker participation and efficacy.

Further, the parties occupying work places should develop more solidarity. The hostilities and conflicts between labor and management inherent in the capitalist system retard worker happiness and productivity. Solidarity means that workers and management together assume responsibility for job security and profitability, included in which is profit-sharing. Both workers and management would benefit from a system that stressed the benefits of solidarity, not the hostilities that result from impersonal downsizing. Perhaps when a company must trim its expenses, *all* workers and *all* management could take a temporary pay cut until the market cycle swings back up. Solidarity thus enables workers to continue to be productive in spite of market cycles.

This working community would also narrow the pay differential between management and labor so as to reduce tensions and provide for a more team-centered mentality. The pay differential between the highest paid member and the lowest paid member should not be more than seven to one, argues Schumacher.[36] That is, the CEO of any company could only make seven times more than the lowest-paid employee. Salary solidarity would produce a great deal of positive morale and thus benefit productivity.

Society's view of the importance of profits must change if technology is to change. The traditional and chief reason for investing in ever-greater amounts of technology is to lower overhead, raise production, and thus make greater profits. Schumacher reverses this strategy. He invests in smaller technology, reaps smaller profits, and isn't supremely interested in maximizing his profits. In fact, he argues that 60 percent of profits should go to taxes and to reinvestment, which would include research and development. He invests the remaining 40 percent by giving 20 percent to various employee projects and 20 percent to worthwhile external and community projects. He claims that this approach reduces the need for social welfare and keeps his business competitive. Newer alternative technologies are developed through his Research and Development department, which also focus on a more humane working environment.

In conclusion, Schumacher's layered view of reality helps him to critique traditional technology and begin to offer an alternative vision of how and why technology should be developed. Perhaps the most impressive element of his thought is the spawning of an international web of practical alternatives. His original view of technology can lead to a different view of economics and economic bureaucracies.

Egbert Schuurman

Egbert Schuurman is a trained engineer, philosopher, and part-time member of the Dutch Parliament.[37] He is also an active representative of an intellectual tradition that has founded the Free University, a major European center of learning. His international recognition suggests the weight of his thought. Though cosmopolitan in his professional interests, this survey of his thought will focus on his major philosophical work, *Technology and the Future: A Philosophical Challenge.*

This book represents his most systematic and exhaustive philosophical treatment of technology. It is an analysis of two schools of thought evident in modern technology. First, there are "transcendentalists," who view technology as a threat to modern freedom. They have a pessimistic view of the place and the effect of modern technology. According to the transcendentalists, technology is like a modern Frankenstein, out to rob us of our essential nature or our freedom. Second are the positivists, who hold the opposite view. For these enthusiasts modern technology is a savior. It frees us from all kinds of obstacles, thus enhancing our human autonomy.

Autonomy can be enhanced, ironically, through the control secured by modern technology. This position is optimistic about the progress engendered by technology *and* about the ability of human autonomy to create near utopias on Earth. Schuurman notes that both positions carry shared commitments to autonomy.

Schuurman's unique contribution to this work is his structural analysis of the meaning and content of modern technology. By "structural" he means a pattern, order, or law for the creation. Structures form the basis for the systematic investigation and practice of technology. A structural analysis attempts to locate the patterns of regularity, identity, and meaning.

The law for the entire creation provides the meaning and the purpose for all of life. It is our task as creatures to listen to this law, with "listening" here meaning a dutiful obedience. Undoubtedly this way of framing the start of philosophical investigation places Schuurman not only within a theistic framework, but a reformed Protestant one as well.

Schuurman makes no apologies for his efforts at a rigorous structural analysis of technology. His task as a systematic thinker arises from his understanding of God's ordering of a meaningful and ordered world. His philosophical reflection on the roots and contours of technology is one means of obedience to God and service to his neighbor. Professionally, reflection represents a systematic way to understand the impact and the meaning of modern technology on culture.

The proper task of general philosophy is to express the unity in the diversity of total reality. By neglecting technology, general philosophy failed to do justice to the diversity of reality and thereby damaged its own insight into the whole of reality.[38] What he means by unity in diversity I will plumb in subsequent analysis. However, at this point we note that he wants to relate his systematic understanding of technology to a more comprehensive worldview.

Schuurman believes that the law for creation forms the structure for all of life, again with structure meaning lawful order, pattern, or regularity. Structure gives meaning and purpose to life. Structure sets the conditions for life. Thus, structural analysis plumbs the conditions for technology, which is to say attempts to understand the place, meaning, and ontological make-up of modern technology. This structural analysis leads him to conclude, for example, that technology is one (and only one) area or part of reality. As such it must take its place among and in the midst of other areas of life.

The blended harmony presupposed by structural analysis is given by God originally to a non-fallen world. We do not live now in that kind of a world. Sin or autonomy grips the creation causing many distortions. One such distortion is the pretension that autonomy brings: something or some process in the creation is thought to provide most, if not all, of life's desires and needs. This process of the exaggeration of one area of life—the technical—represents what Schuurman calls absolutization. When one area of life is exaggerated, the rest of life is reduced or undervalued. This shrinking process is called reductionism. Redemption or wholeness may be evidenced in life when we see a restoration of more fulness and harmony.

How might a structural analysis of modern technology begin? Technology has meaning or purpose because its activity is part of a grand meaning given to the creation. This grand meaning is made possible by a structured creation. This structured creation carries with it principles or norms that particularize instruction in requisite aspects. As part of our identity, humans must give form and shape to the creation according to these pre-established norms or principles toward some human end. Thus, there is an explicit teleology or end view in mind in technology, according to Schuurman.

Further, technology, like science, is interested in abstracting problems and activities from the matrix of life for the purpose of study. Through analysis and abstraction the engineer breaks down problems into universal component problems to arrive at universal component solutions. Interchangeable parts are an example.

The engineer is interested in lasting universal coherent knowledge, just like his scientific counterpart, though in the technical aspect. Modern technology especially is designed to give form to projects from a distance—that is, through the means of technological operators or more or less automated machines. This distanced relationship differs greatly from classic technology where humans had a more direct contact with technological projects. Further, the degree of freedom afforded in the preparation and the design of projects was greater in the classical period. A structural analysis of modern technology reveals an immense contribution by the focused, analytical, segregating methodology of modern science. Nevertheless, argues Schuurman, the practice of technology is tied still to the matrix of meaning we call the creation.

To do structural analysis is to inquire into the areas of fundamental importance not only for technology but for all of life as well. Structural analysis is fundamentally important because structures form the environment for meaning and purpose. Meaning and purpose make technology what it is; it receives its identity from structures. Thus, to do structural analysis is to inquire into the foundational conditions for the identity and the conditions and meaning for technology.

As I have just noted, Schuurman claims that it is impossible to contemplate the meaning of modern technology apart from the contribution made by modern science. Science is the necessary systematic study of the order and regularity for reality. Technological science also is interested in the systematic study of any given set of conditions but to different ends or goals than science. The goals of the technologist are more practical and project oriented. These goals require the ability to fabricate or form and produce material reality into a coherent technical project. This powerful ability to manipulate nature for predetermined technical ends represents for Schuurman the central defining feature of technology.

Modern technology enables the technologist to gain control of the forming process from a distance through the use of what Schuurman calls technological operators or automated technological products. Further, through scientific analysis and abstraction the engineer breaks problems down into component parts to the end that a solution or a goal may be reached. Sadly, the worker is part of the automated process that engineers seek to control from a distance. Control is considered necessary because assembly-line workers are thought to be an extension of the machine. It is in this systematic breakdown of reality for specialized study that science aids technology but hurts the modern worker.

Just as Schuurman is not antitechnology, he is not antiscience. Science has made a valuable contribution to the history of technology.

The systematic, orderly analysis that defines a significant portion of modern science has provided technology with a helpful methodology as well. When technical analysis dissects reality into parts in order to find solutions to problems, a skill engineering is based on, we all benefit, argues Schuurman.[39]

The fundamental problem of both science and technology is the lack of coherence or integration because of the fragmentation that results from a lack of understanding of the whole of reality. Therefore, Schuurman says that any technologically fabricated product or project only forms an *analogy* to the coherence found in reality and to the coherence of scientific knowledge. The reader is advised to keep this notion of coherence in mind.

The structures that present themselves to the philosopher remain in spite of reductions, absolutizations, and (to use the language of Ellul) our idolatries. Something of the meaning and the good purpose of technology remains in spite of us, claims Schuurman: evil is not sovereign. Thus, there can be no technological determinism, no "technological society," or fundamental dialectic to life. This view of the reality of structures, therefore, separates him from pessimists.

His view of structures, further, separates him from optimists. Accordingly, Schuurman sees the distortions affected upon the structures by human arrogance; he is not naive about evil or ignorance. Technology can and does become distorted because human actions are sometimes twisted and distorted. Further, though technology should exercise a *limited* influence in all areas of life, we must not think that technology alone or primarily can solve all especially nontechnical problems. To think so is to lead to technicism or the exaggeration of the place of technology within our lives. Instead a variety of principles and areas of life should be harmoniously blended to solve difficult problems.

He is not a realist because his methodology in not guided by utilitarianism. Schuurman does not believe that humans are utility maximizing, rational creatures who trade off bits of nontechnical reality for more or less of technology. He has affirmed the rich methodology used by realism to evaluate the effects of technology, but he could not support the economic materialism that undergirds realism.

Nor are Schuurman's views simply a more complex version of those of Schumacher's. Schuurman's views come as the result of much more intellectually rigorous process and complex ontology, their similarities notwithstanding. Schuurman, unlike Schumacher, is struggling with the relationship of "big" to "small" technology and the cultures that are related to both. The following statement betrays both this rigor and his integral view of reality:

> In the first place, it is necessary to satisfy the *cultural norm* of differentiation and integration, of continuity and discontinuity, of large scale forming and small-scale forming, of uniformity and pluriformity. These various components must not be regarded as contradictions.[40]

The term *norm* deserves special attention. A norm is a principle, a command, a must for human activity. A norm gives identity to different kinds of technological projects. Small- and large-scale technologies are not so because of the genius of humans. Schuurman is arguing that there exists a normative—an imperative—that small and large technologies must exist to complement each other. In a complementary relationship, a harmony or "salvation" can exist.

I return to Schuurman's definition of technology to analyze his understanding of the relationship of human identity to tools. He adds to his definition by saying that technology is "the activity by which people give form to nature for human ends, with the aid of tools."[41] By "form" he means giving shape or visible expression to a technological object; he does not mean an Aristotelian rational essence, nor is the law to be conceived in this manner. Rather, giving form to something means shaping it into a desired end or forming a technological object with some direction in mind. Ends and means must be informed by the lawful structures for life. One can cut down a tree and form tables and chairs, but this forming only means that one takes the potentialities inherent in materials and brings them to reality. The limits of the materials, purpose of the project, and the forming process itself serve to limit the identity of the project at hand. Thus, the structures for life inform the identity of a project even before humans touch the original material. Human identity is found partially in forming, shaping, and developing reality, not creating or dominating it. The crucial difference between these two stances is one between a creature whose existence is given and an alleged creator whose existence is believed to be formed through the might of autonomy.

The ability to shape nature is given to humans as part of our identity, argues Schuurman. However, while we must shape nature, it is destructive to be autonomous. This myopic attitude leads to a distortion of reality. When the creation is manipulated in distorted ways, as it is when humans have to mimic the motions of a machine, technology distorts our character. At the same time, Schuurman would insist that forming is, per se, a God-given, necessary human activity that should not and will not fade from human use.

Through a complicated but brilliant analysis of the structure for modern technology, Schuurman shows how modern tools have surpassed in reliability and accuracy the proficiency of human counterparts. The machines of the Industrial Revolution greatly exceeded the reliability and power of human muscles. The modern computer vastly outstrips the human brain in speed of mathematical calculation. Thus modern technology *objectifies*—or makes real, external, and independent—many human functions in a manner that greatly exceeds human capabilities. This Promethean ability of the modern machine to exceed human abilities leads humans to desire more technology until addiction threatens our humanity.

It is impossible for technology to be neutral or value-free, argues Schuurman, because of the structural conditions set for technology. Humans design, implement, and complete technological objects or projects according to some design philosophy or view. This view, in turn, is based on the structures for reality. Quality engineering presupposes a thorough knowledge of materials being used. Materials, in turn, have an identity based on a structural identity. Hence, material use is based on the meaning given to properties. Indeed the term *properties* connotes a law for creation. Thus, the standard engineering term—*proper*—used normatively to describe a standard for material use describes the structural conditions for reality. Hence the word *improper* denotes a violation of the structures for creation. Improper use of materials can lead to ethical and legal problems. It may be concluded, therefore, that the basis of ethical, legal, and professional responsibility must be located in the structural conditions for technical development.

Further, because the structures exist *for* reality, they have a determinative meaning for life. The law of gravity significantly determines the aerodynamics of an airplane. Indeed, wings are designed with aerodynamics and the law that stands behind it in mind. Thus, Schuurman is arguing that autonomous technology is a myth because the structures for reality give a meaning to forming before, during, and after human interaction. There cannot and may not be, therefore, autonomous technology because technology is never a law unto itself.

Returning to the theme of abstraction, one may say that its intellectual twin specialization has a downside, especially in engineering. Focusing on isolated, fragmented reality without a view of the whole often leads one to conclude that an entire worldview should consist of that one sliver of reality. Engineering schools turn out brilliant technical engineers who, nevertheless, have lost sight of the relationship between their narrow skills and a larger view of engineering's relationship to reality as a whole.[42] He thus concludes his analysis of the foundations for reality. He then moves on, in *Technology and the Future*, to analyze the dominant thinkers in the philosophy of technology.

Schuurman again locates two groups of thinkers in the field of the philosophy of technology: transcendentalists and positivists.[43] Transcendentalists are so named because they take as their starting point or supreme principle the alleged autonomy of the human person who is thought to stand behind or above the grip of technology. Hence, the person is thought to transcend technically dominated reality. Accordingly, a person must be free from external restraint at all costs because to restrict the person technically is to rob one of an essential nature. Human freedom is the essential core of human identity. Thus, they start with reality inside the human because freedom originates within the human mind. It is believed that a technologically determined society threatens to undo our essential nature. Jacque Ellul is a transcendentalist, according to Schuurman.[44] Most transcendentalists think that modern technology represents the greatest threat to human freedom because it restricts rather than promotes freedom.

Many transcendentalists are nostalgic for a return to a time before the Industrial Revolution: a time before experts, and a time when human action was not emasculated by technology. According to the transcendentalists, modern technology eventuates inevitably in a total technocracy in which the individual is robbed of freedom and transmuted into a manipulated mass-person.[45]

Because transcendentalists hold autonomy as the supreme virtue, they fear the "technocracy." They fear that a social rule will amorphously but lethally eventuate from an elite, nonelected group of technical experts who will surely destroy society.

Transcendentalists sound additional alarms. Technology feeds upon human and subhuman nature. Species die, humans yearn for freedom, and parts of our environment deteriorate. Wholeness or completeness cannot be found, argue transcendentalists, in more technology. Technology can only fragment and erode the quality of life. Further, in our rational-instrumentalistic lust to dominate nature, we have lost our rationality and our freedom. We live in a mythical world, if we think we control reality, because our control of reality is artificial or partial. Transcendentalists rest their case.

Positivists start, on the other hand, with reality *outside* the human. This means that they look at the positive, objective technological facts or objects at hand or near to our gaze. Technical facts always and inevitably lead to progress because technology always leads to progress. Again, progress is the equation of total human betterment with perceived betterment in technical development. Rather than being a menace to human freedom, technological development provides the essential conditions for human freedom. We are freer because the machines of the Industrial Revolution delivered us from difficult, meaningless work to pursue our own agenda. Positivists form the largest block of people committed to the kind of optimism outlined in chapter 1.

Positivists remain optimistic about the future because of what technology has accomplished in the past. The positivist argues that society in the future should be governed (an ethical imperative) by a technical elite who benevolently will distribute the benefits of an essentially good and wholesome technology. This optimism ultimately, and hence religiously, is rooted in a trust in instrumental or means-oriented rationality that seeks to orchestrate reality according to technical norms. Hence, positivism is optimistic about the human ability to correct problems through technology and because humans are essentially good, moral, and rational.

Schuurman's most creative contribution to the philosophy of technology is his analysis of the deeper "religious" affinities between positivism and transcendentalism. Both movements stress the autonomous use of and dependence on, say, information as the key ingredient for knowledge. That is, the "information age," so named because of the commanding presence of various kinds of information, depends on autonomous individuals who maximize their freedom through the collection of an ever-increasing amount of information.[46] It is not ironic, therefore, that Ellul retreats into an intrapersonal world of communication as his only alternative to modern technology.

Both groups believe that society represents a system of interactions held together by communication. Development of an information society is essential to our nature because information is the key ingredient needed to maximize our freedom. Modern freedom is enhanced, according to both camps, by first stockpiling and then applying knowledge to society. This belief states that progress will arise inevitably.

This heavy emphasis on information as the conduit for social progress leads to a reductionism, argues Schuurman. Reductionism is the shrinking of reality to one or a very few aspects such that society's meaning is thought to originate from the free application of knowledge. Reductionism results from an **absolutization** or an aggrandizement of social communication.

Thus, while transcendentalists and positivists *appear* to disagree, at a deeper religious level however they agree that human beings are autonomous and that their autonomy is best expressed through the increased commitment to information.[47] Schuurman argues that this debate is a "family quarrel" between different people who share a common reverence for the free autonomous personality. Whatever may be their ideological differences, such diverse people as the Protestant Ellul and the atheistic Marx are secular in their technological outlook because they share no vertical or God-directed orientation in their technological reflection.

Further, nature is seen in both camps as an arena in which to demonstrate human freedom. This view motivates the optimists to dominate nature. Finally, both positions share the belief that human instrumental rationality is capable of mastering the world and guaranteeing meaning. Schuurman would argue that meaning is discovered, not generated.

The meaning, purpose, and order for technology cannot be found in human autonomy or in more technology, argues Schuurman.[48] He believes that God gives meaning to reality. The job of technology is to fabricate or give shape and form to nascent meaning. Thus, a tree may be cut down and shaped to meet our legitimate needs. This giving shape can only take place within the reality presented to the engineer, not by a reality he creates.

Humans have become enslaved to technique. In this Schuurman and Ellul agree. However, unlike Ellul, Schuurman believes that our autonomy is a pretension or an arrogant, make-believe illusion that we hope will save us in our attempts to remake the world after our own technological image. Our autonomy turns on us to tyrannize us when life's rich possibilities shrink because of our commitment to technological autonomy. The shrinking integrity of the natural environment is a case in point. Thus, at the deepest religious level, Schuurman is neither a pessimist nor an optimist because of his rejection of autonomy as a starting point for life.

THE BASIS FOR LIBERATION

Schuurman takes his starting point in the word of God.[49] This starting point represents a reformed Christian faith.[50] God's word, or the effectual revelation that governs reality, is the express means for God's care of the creation. Because it exists external to humanity and serves to direct or guide our behavior, it is heteronomous in character. That is, Schuurman takes as his starting point not autonomy but heteronomy. It forms the context for freedom. Freedom is enhanced by making God's precepts the basis for our lives. Technology enhances our freedom when the precepts of God become our operative principles. These precepts form the basis for the structures that give shape and hue to life. The principle of stewardship can be used as an example. We use technology wisely when we steward our environment in such a way that nature is simultaneously cared for and developed. Thus, Schuurman uses the norm of stewardship to talk about agriculture:

> Taking the way of normative structure begins with acceptance of the motive . . . from which agriculture must be carried on, namely, that of harvesting, keeping, and maintaining. Such "keeping and maintaining" extends to preserving and improving the productivity of the soil and conserving the biosphere, for example. . . . The perception that agriculture ought to fit the given situation in which nature, environment, and landscape are found has been too little present.[51]

Such principles or norms can be found for all aspects of life. Further, in spite of all of the distortions, idolatries, and mistakes, something of the very good character of technology remains in our world.[52]

God's word or precept for creation gives structure to human activity. We ignore it to our common detriment. When we turn from our agenda to God's we find that creation can be *opened up*. By opening up the creation, Schuurman means to say that humans have the ability to *disclose* or make apparent the practical meaning inchoate in creation. Technology, therefore, helps us realize possibilities nascent in the creation, regardless of the religious condition of the person. Creation's powerful and dynamic structure forces humans to continue being technological because failure to do so would violate part of our nature. That is, ceasing to be technological would be like not eating; we can't stop for long because life demands it. "Whenever humanity subjects itself in the present to the meaning dynamis, the future opens itself as progression."[53]

It should be noted that Schuurman claims to be a progressive, by which he means to say that his fundamental ideological posture in life is to be expectantly awaiting exciting new possibilities for life. That is, he is not fundamentally a conservative or a liberal but attempts a creative, third way beyond most problems as they are stated.

Our decision for or against God affects our technologies. Humanity is subjected to God in being technical. Hence, one of our essential human conditions is to be technological. Thus, when we act technologically properly, we enhance our humanity. The distortion comes not in being technological, but in the distorted manner of technological development. Sin enters the picture by the way we distort our technological tasks in life. Distortion primarily occurs when we attempt to rip technology out of its given coherence. That is, when we absolutize or exaggerate the place and the importance of technology, to the detriment of the rest of life, we lead ourselves and our technology into distortion.[54] When our technology distorts or erodes (nature for example), exaggeration lurks.

More must be said about the relationship between norm and principle to the structure for creation. The principles not only form the structure or meaning for creation, they form the dynamic for creation. Dynamic means the motive force by which the entire creation is driven. The motor, or force, behind this power is God's word, or, to use the theological term, general revelation. Revelation is given in wholeness and thus forms an integral web of meaning for life. Thus, reality is interdependent in nature. Only in sin or because of science is reality torn apart.[55]

Any science or field of investigation necessarily abstracts one area of creation from the whole and especially focuses on it.[56] If a discerning posture is taken by the scientist, then scientific discoveries can lead to wisdom or knowledge suited to all of life. However, if, for example, instrumental reason attempts to control and dominate the DNA make-up of a person without a view of the entire DNA field or a view of the child as a whole person, then distortion lurks. The distortion reduces the quality of life by treating life as "pieces" to be combined at will.

This motive force is clearly seen in the particular principles that form the structure for life. According to Schuurman, the following principles are relevant for given "rooms" or areas of life. These principles or norms provide an integral, normative structure for creation: centralization/decentralization for history; clarity for the lingual area of life; interaction for the social area of life; efficiency for the economic area of life; justice for the political area; stewardship for the biological area; caring and loving for the "ethical" or family; and certainty for the faith area of life.[57]

It is important to note that Schuurman believes that these norms apply to all people and that they are followed or spurned on the basis of one's religious or fundamental predispositions. Further, while particular principles apply especially to one area of life, analogous principles can be found in all other areas of life. This is so because reality is integral in its constitution. For example, although stewardship allegedly is the principle for the economic room, we can see that for Schuurman there are analogies of stewardship in biological and psychological life. All analogies, however, share the same core meaning: a conserving, preserving, enhancing of creation. We preserve the creation both when we rest and when we add nutrients to depleted soil.

These principles can be discovered through scientific investigation and formulated into laws or regularities that apply to a given discipline. If one detects that a regulative principle reoccurs and serves to give identity and structure to one's field, Schuurman would argue that a norm or a standard has emerged. The grand norm for technology is the ability to give form to inchoate reality, where meaning is only implicit to the technologist. Engineering is established on this fact.

Norms give technology meaning and form the basis for the responsible use of technology. Technology is used properly when it cares for the environment, promotes family commitment, increases justice and democracy, enhances education, and so on.

Responsibility is located in revelation for Schuurman; its origin is not human Reason. This ever-present word of God calls forth a response from all human beings (assuming, of course, that animals are not technological). Hearing and doing the word of God becomes the basis for freedom and thus the responsible use of technology. Freedom comes from a multifaceted proper response to God's precepts. This freedom is limited by the structural nature of reality. Thus, an area of life like technology can and should be viewed as deeply meaningful, but its effects must be limited.

This view of responsibility and its relationship to structures delivers Schuurman in principle from optimism and from pessimism. Technology is dependant upon God's sovereign word. That word cannot be resisted absolutely by humans. Hence, Schuurman is not a pessimist because man is capable as a responsible person of a genuine response. Nor can technology control society, though it does have entirely too much influence, because the word of God and not of man is sovereign. On the other hand, because this word serves to direct and to limit technology, its fruits cannot possibly produce a universal salve for life's serious problems because reality is multifaceted or many-sided. Only by the false faith in the innate goodness of humanity and in the autonomy of instrumental rationality could such a thing be believed; so Schuurman argues.

This does not mean that the pretension to the autonomy of instrumental rationality does not wreck havoc in our lives. Technicism, or the absolutization of technique, is the sin resulting from autonomy. These exaggerations do wreck havoc.

Having granted the differences between Schuurman and Ellul, I push on to view the similarities. Schuurman would want to affirm many of the insights Ellul puts forth on the nature of technicism because of the obvious imperial nature of modern technology.

Technicism reflects a fundamental attitude that seeks to control reality and resolve all problems that arise with the use of scientific-technological tools.[58] Hunger will not be resolved merely or primarily by technical means. The origins to this problem spring from the distortions plaguing the human condition.

Technicism becomes an exaggeration because of the lust to control reality. This lust for control of reality comes from autonomy and manifests itself in the attempt to break down reality into its simplest building blocks. Once subdivided, then conquered, we next attempt to reconstruct reality using our autonomous instrumental reason. Frederick Taylor's system of scientific management is perhaps the best example of this lust of instrumental reason. Accordingly, the effort of the laborer is broken down into its most repeatable, controllable, simplest movements. The shop foreman using this knowledge could control the laborer by redesigning the work and by incentive. Work was to be the endless series of simple, repeatable motions. When human movement was controlled, then its lucrative products came more efficiently, thus affording the worker more salary. Worker opposition to "Taylorism" soon faded because of the narcotic effect of money.

Schuurman calls technicism a "moloch" or a false god. A false god means the same thing as technicism. At the same time, Schuurman will not view technological practice as a value-neutral means (or "*middel*," to use the Dutch term) to some predetermined end. This view is taken generally by optimists. Realists give it yet another shade of meaning because of the way they define instrumental rationality as a neutral means to a socially desirable end. Yet Schuurman objects when he says, "There can be no neutral or value-free technology because structures give meaning to reality."[59]

Schuurman's notion of responsibility suggests a practical direction. He favors the integration of small-scale, labor-intensive, and environmentally conserving technology with our current, large-scale, capital-intensive kind of technology. The precept of historical diversity informs the large/small-scale notion. The norm of the goodness of labor defines labor-intensity. Stewardship is the norm for environmental care. This alternative view of technology is meant to complement not dominate other areas or aspects of life. Thus, he shares with Schumacher a view about what place and location modern technique should occupy because of his view of the structured nature of reality. It should complement, not dominate, life.

Further, technique should be durable, energy saving, and aesthetically pleasing because the principles of stewardship and aesthetic harmony demand it. Finally, we need to develop a national energy and technology policy stressing how technology complements life, because the norm of national justice demands it. Justice for Schuurman means that all life communities have a right to protection and promotion by the state. Debate leading to this national technology policy should allow us to adjudicate how *all* communities within the state should benefit from such a policy. This necessary cultural debate would serve to slow down technological innovation so that multidisciplinary and public reflection could match technical innovation, necessary if we are to live responsibly.[60] Had we taken our time with fission, we may have avoided the difficulty surrounding waste treatment. Thus, Schuurman argues, "engineers should not strive to do all things possible, they should strive to do all things necessary."[61]

Technicism never can lead us to responsibility, claims Schuurman; it can only threaten us. The nuclear arms race is a case in point. Believing that we can find safety in ever-greater amounts of armaments, countries of the world spend trillions of dollars each year in making and marketing weapons. Yet the use of these weapons, especially the nuclear kind, brings untold havoc and threatens countless lives each year, claims Schuurman. Ironically, though the state is called on to do justice through the way in which it defends its citizens against aggression, there can be no significant defense against a surprise nuclear strike. Thus, Schuurmann claims, the "Star Wars" missile defense system represents yet another optimistic myth foisted on the taxpaying public, claims Schuurman.

Schuurman applies his notion of responsibility to the use of the computer. In keeping with his understanding of the historical norm of differentiation, he argues that our culture is becoming too centralized and that we are in danger of losing our freedom. Responsible computer use begins when the computer is used for more decentralized, confidential decision making. The public must be protected from some unscrupulous people who would use stored information without a person's consent and thus invade one's privacy. Privacy is here the freedom he seeks to protect.[62] Further, computers can aid deficient human functions. Such is the case in computer-assisted wheelchairs.

Schuurman also tackles the technological side of genetic engineering. Generally speaking, Schuurman is in favor of genetic therapy so long as the procedure is meant to correct a specific medical and/or medically diagnosed problem. He rejects generalized genetic engineering and many forms of cross-species manipulation and implantation because these procedures reduce humans to bundles of technological genetic fodder used for manipulation and control. Treating humans accordingly reduces the meaning of our humanity to that of a genetic puzzle, the secrets of which are to be unraveled by the expert. Thus, human dignity is reduced. This view cannot be far from the equally myopic view that believes that people are just complex information processing systems. Humans are not just so much "stuff" to be manipulated or arranged to fit a "better" definition of humanness.[63]

THE PLACE OF TECHNOLOGY

I believe structuralists offer a unique vantage point from which to view the place or the location of technology within life. Their program for the locating technology rests upon the idea that life consists of many aspects or layers that are, nevertheless, intertwined at the core. These aspects are structural wholes that permeate reality and thereby give it meaning and limits. That is, each "room" has set for it conditions that give it identity. Technology is one room with its unique but limited identity. This identity can not be totally obliterated nor can it obliterate all other rooms. Its place can and indeed is greatly exaggerated in our culture. This state of affairs is called technicism by Schuurman and is a situation to be avoided.

However, technicism can and is mitigated by the nature of a diverse reality. Diversity forces on us a view of the contexuality of technology. That is, technology must take its place among a variety of different rooms. Each room should complement, not dominate, life. This leads the structuralist to extol an integral view of reality, a view where aspects and their attendant principles harmonize with each other. This view of harmonization is behind Schumacher's notion that small is beautiful. Thus, Schumacher argues that technology must fit and enhance the actual size of humanity; we must not adapt ourselves to large technology to the extent optimism demands. Technology must be adapted to our needs. Harmony is not something added to life as if harmony were a necessary appendage to life. Rather, life at its core is harmonious for the structuralist. That technology too often fragments life is regrettable and perhaps the major problem in the twentieth century, but it is not totalitarian, nor is fragmentation absolute.

The strength of their convictions—and this is more true for Schuurman—rests on a belief in God's providence. This notion stresses the sovereignty of God, not the sovereignty of evil, à la Ellul. Rooms receive and maintain their identities because God gives normative commands first, then identity to areas of life, so that they may be preserved and maintained. These commands provide not only the basis for freedom, but they resist any totalitarian attempt at sovereign control. The fall of the former Soviet Union gives ample evidence to this claim that no human power is sovereign. If any nation or person unwarrentedly assumes sovereignty, reality, that is God's power, will resist. We have freedom but only within the limits of law. We may arrange the furniture to our favor and liking, but we may not completely obliterate the Architect's design and purpose for the room.

I believe the structuralist position offers several advantages over former positions. Rooms are not balanced, as per realism, through some kind of rational tradeoff. There are no utilities to be traded off, as if happiness comes in individualistic, materialistic bits. Neither do structuralists offer inherent contradictions like Ellul. They offer inherent wholeness and integrality. There are principles or precepts that can complement each other and the rest of reality. These precepts enhance the quality of our ethical lives. If wholeness is a virtue for humans, why should it not apply to the rest of reality? That conflicts between rooms do occur is all too real. However, conflicts signal a potential problem and, therefore, provide an opportunity to assert a solution.

Nor is the technical room the grand room as optimists would have us believe. This reduction of the meaning of other rooms to its technical adaptability represents a truncation of reality and a myopic view of life. Good technique complements the rest of life and is adapted to the actual size of humanity. Good technology enhances the potentials built into the rest of creation. This is not a mere slogan, as the alternatives spawned by structuralists show. Structuralists are not sanguine about the possibility of real "salvation" or harmonization because neither correction nor responsibility ultimately resides in their hands. Structuralists believe freedom originates from the apparent irony of giving up ultimate control of one's life. When one listens rather than commands, or tries to control reality, one has become more responsible at that moment. This universal need for responsible freedom characterizes a human journey that has been crafted by God's whole-life directives.

APPRECIATIVE EVALUATION

What can be said of structuralists? Have they helped us to learn more about the place and importance that modern technology should and does occupy in our modern lives? I think they do, though I say this with some hesitation. I believe this position is the most credible and potentially the most fruitful one developed to date. This confidence is based on several strengths.

Structuralists escape pessimism's root problem of absolutization of technological evil by linking the inherent goodness of God with His creation generally and technology specifically. No amount of human evil can completely and totally control all areas of life. The parental phrase directed at children to "turn off the TV" means that the needs of family and school occasionally take precedence over technology. Because other areas of life cry out, society cannot become a "technological society." It is surprising that Ellul did not take such a path, given his reformed Protestant faith. I suspect that Ellul did not take a more positive path because of his inclusion of Marxism into the other half of his worldview.

Linking God's revelation to structures strengthens and makes furtive Schuurman's effort to reflect systematically upon the meaning, limits, and function of modern technology. Without this systematic view of structures, locating the matrix of life and understanding technology could not be attempted. Therefore, while secular thinkers may disparage or ignore the theistic thinking, I can find no one who matches Schuurman's systematic investigation into the ontological conditions for technology. This rigor enables him to articulate clearly the meaning and the limits of modern technology. This view of a structure, then, provides the basis for locating and contextualizing technology within the warp and woof of other equally valid areas or aspects of life. Further, it enables him to discuss the conditions in which technology specifically, and life more generally, may be understood. A universe of precepts does not represent a loss of freedom; it represents a world of cohesion and meaning in the midst of fragmentation.

Schuurman's view of structures undercuts several additional problems before us. First, optimism is inherently wrong if it thinks there can be a technical solution to every problem. The multidimensional, many-sided nature of reality demands that a problem as complex as hunger, for example, be tackled by a variety of principles and disciplines working together to begin to craft a solution. Pessimists could greatly strengthen their case if they understood the multi-dimensional goodness of life. Ellul's expanded awareness of the overbearing nature of modern technology as well as his writing and mass production of books testifies at least to his conscious escape from "the technological society." Because he in his consciousness has escaped the technological society, he is free to direct portions of the economic, technological, educational, logical, confessional (or faith), and symbolic aspects of life to serve the concerns of freedom.

Further, realism rightly stresses that a variety of principles should inform technological choice. However, because they have no differentiated or coherent yet multilayered view of a context out of which these principles arise, they cannot do justice to exactly *how* and *why* one is to make a tradeoff. The primary principle for determining how much of "X" must be surrendered so that more of "Y" may be obtained is economic rationality or utility maximization. Realism gives us denuded principles in a world without context. Structuralists do not have this reductionistic problem because their layered view of reality demands a context rich with meaning and harmony of principles. Therefore, their layered view of reality leads them to argue correctly that modern technology represents an idolatry or an absolutization. Schuurman and Schumacher's views about the many-sided nature of reality cannot be dismissed, therefore, as "misplaced theology" because their view of reality is too complex to be the product of merely one discipline. The true definition of the origin of freedom and the nature of reality represents the true stakes in the debate between the theist and the nontheist. Structuralists argue that the first three positions studied lead to oppression, not to freedom. Freedom only comes when we listen to the voice of Another.

Structuralists present us with broad, somewhat flexible principles that may be used for technological evaluation. Accordingly, technology may be placed *within* not *over* life because reality is diverse yet coherent at its core. In other words, technology, like one room of a home, is placed in relationship to other rooms. Their identities and purposes are distinctly crafted by the architect, yet the crafting is done with an integral view to the entire home. Principles define aspects; both are meant to complement each other. The design of automobiles can be aesthetically pleasing, safe, family-oriented, protective of the environment, and just in their use of space and limited resources. With the increasing use of the electric car, multi-faceted responsibility will become even more of a reality.

The rich view of life applies to our understanding of human nature as well. Structuralists do not fall prey to realism's seemingly simplistic utilitarian view of the person: utility-maximizing; pleasure-seeking; and rationalistic. Their view of the person argues for a holistic understanding of the person whose nature is severely violated when reduced to a utility-maximizing, rational machine. In short, structuralists have advanced the discussion about the location and import that modern technology should occupy in our lives because they have deepened and nuanced our understanding of life.

The structuralist's analysis raises grave questions about the relative shallowness of the realist's methodology. A pleasure-seeking, rational, utility-maximizing person who forms the basis for the realist's view of risk analysis and tradeoffs can only lead to a superficial technological evaluation. If reality is more complex than realism accounts for, then their methodology is superficial in principle; that is, in view of the nature of reality and humans who inhabit it.

Further, if reality is interdependent and multifaceted as structuralists argue, then locating the study of technology within a matrix of disciplines, themselves the products of the matrix of reality, seems wise. This interdisciplinary approach more effectively addresses persistent problems like hunger, poverty, and environmental degradation. A coherent interdisciplinary approach to a problem strengthens our collective and personal responsibility because multifaceted cooperation is encouraged by the nature of reality and the methodology employed. Two, or in this case multiple, disciplinary heads being better than one seems oddly appropriate here. Further, this same interdisciplinary method becomes the basis for a liberating methodology for technical evaluation because principles, or norms, are built into life. Thus, principled ethical evaluation is not a superficial addition to an otherwise value-free discipline. Rather, norms or principles form the foundation for this view of reality and thus their study takes us to the roots of life.

Therefore, this philosophical method leads us to the rigorous questioning of foundational assumptions. A foundational assumption is a
working idea with an operative force that touches all relevant theory
and practice. All positions surveyed up to the structuralists have held
to autonomy. The very notion of a *foundational* assumption implies a
coherent view of the relationship of assumptions that occur before
theory to the act of theorizing and this complex act, in turn, to the
practice of technology.

Further, when structuralists trace the effects of modern technology
into natural and cultural systems, they provide a substantial critique
against the overidentification of culture and nature with technology.
When technology itself becomes culture and a barrier to nature, reality is violated. Urban technological jungles result. This reality does
threaten to become a "technological society." Adjectives used to
define society tell us that technology has not been integrated with society and nature; rather, it dominates them. A case in point is the net
qualitative and quantitative effects of the modern automobile. When
this one form of transportation cannot take its place alongside many
forms of transportation, congestion, or a lack of harmony and integration, exists, as does gridlock.

Finally, structuralists have encouraged practical alternatives to
problematic technologies, though I say this with some pause. A
multifaceted approach to life and to study lends itself to technological diversification much more so than does the market. The market
was designed to weigh *only* economic alternatives.[64] The relatively
limited efficiency principle that guides the market cannot begin to
match the ethical diversity apparent in the structuralist's approach.
Technological diversity is especially apparent in the work of E. F.
Schumacher. His practical alternatives show how changes in the
nature of technology require changes in the nature of life, indeed in
our deepest religious commitments.

When structuralists call for the depth of change required for a
healthier kind of technology, they manifest perhaps their chief
strength. Structuralists do not want merely a new technology; a technical fix consisting of new gadgets that correct the old ones is not the
primary goal. Technology cannot correct technology. A completely
new and more deeply experienced solution is needed. Structuralists
have advanced the discussion about the place and meaning of technology precisely because they have begun to see the global, integral,
and radical effects of the beliefs and principles that form the foundation for modern technology. A lasting and sufficient change for a

more humane technology must keep this integrality and depth in mind. Changing merely the kinds or the arrangement of tools is superficial and signals a symptom of the problem of technicism because technical changes cannot correct technology.

There is no mere technical fix for a deeply human problem. A culture's tools emerge from and subsequently effect deep commitments in life. Thus, a deep change in life is needed if we are to change our tools significantly. I believe, therefore, that some of the criticism directed against Schuurman and Schumacher originates from a belief that there is nothing fundamentally wrong with technology that a change in tools or the social arrangement of tools can not fix. Hence, those arguing for deep changes are viewed often with mild disdain.

Unfortunately, structuralists' efforts sometimes encourage the disdain. I argue that structuralists are not structural enough in their analysis and remedies—that is, structuralists weaken their view of structures *before* any well-thought-through alternative can be systematically explored. This is ironic since they tout the integrality of reality. First, I will consider Schumacher's work.

Schumacher's view of small or intermediate technology lacks the precise methodology to integrate his view of technology within an existing culture committed to bigness. A First World culture is more structurally diverse and thus more complex than is a Third World culture. Thus, a "small" technology that must do justice to any given historical situation must be engineered to fit a culture.[65] Structural differentiation in the United States is more complex than is the mix of Third World countries and hence a different kind of "small" technology is required.

"Small" implies a relative position that has not been worked out by Schumacher or his disciples. What is small for America or Europe might be large for Zambia. McRobie presents various alternative technologies without any explanation as to how, why, or if mainstream technology is to change or absorb these alternative technologies. It is as if these "intermediate technologies" must exist in some parallel but undefined fashion to the dominant trend of culture. If Schumacher is serious about the problem of "giantism," then a proposed national technology policy that specifies exactly how a variety of technologies should be blended into a coherent, integral whole addresses his concerns.

This policy should not be a government-controlled policy. Both Schumacher and Schuurman would abhor such a plan. It would take a formative institutional presence to coordinate urban, suburban, and rural sector and institutional analyses and implementation. Specific technologies should be matched to fit not only sector needs, but institutional needs as well. For example, a large-scale nuclear reactor located in a center city (sector) servicing only a few families (institutional) would be imprudent. More urban, rural, and possibly suburban windmills, say, used to *complement* existing power sources, especially for the poor, seems needed. Rural/urban, skilled/unskilled, technologists/craftsman, centralist/decentralist, small/large dichotomies will influence families, schools, churches, governments, and business differently depending upon the degree of cultural development. The goal is to promote integrality or cohesion between the poles of the dichotomies. For example, while in Zambia, I learned that simple hand tools were in short supply. A coordinated national policy stressing the creation of hand tools made from labor-intensive machine shops located in strategic village centers could address many problems for the Zambians. Similarly, America's urban infrastructure is badly in need of repair. A labor-intensive policy aimed at employing urban craftsman and many small businesses complementing existing union and government contractors seems plausible. Money in the federal budget and taxes levied for the suburbanite who uses city infrastructure could pay for this work. Putting the poor to work could reduce welfare payments thus causing a cut in our federal taxes.

Because Schumacher does not sufficiently develop his view of exactly how technology can fit into these structures nor into culture, his view of the line between where small ends and large technology begins is not clear. This fact is especially a problem for America given its obsession with economic growth and technological concentration. The economy of the United States often evolves into *economies of scale.*[66] When public utility monopolies dominate an industry, the context and norms for "small" technologies must be sharply drawn. If many firms and kinds of technologies are to compete in a culture, then Schumacher must show how diversity of any kind is not wasteful of resources, a state of affairs he wants to avoid.

The problem of fit extends to his theory of management. Appropriate technologists have moved into host countries with a First World, hierarchical view of management. This view robbed many Third World managers of their dignity. Consequently, local self-reliance, a must for appropriate technologists, waned. Thus, one aspect of the economic sphere—management—weakened entire projects because of the lack of seeing the structural integration of life.[67] In a country that is rooted in the tribal system, a manager who oversees members of several different tribes is perceived to be a petty dictator because only warrior-chieftains who won battles over many tribes ruled hierarchically. Technologies must complement existing cultural patterns.

Thus, appropriate technology must be historically sensitive. Structures are found within the matrix we call history. Failure to allow this reality fully to inform one's analysis forces one into an abstract, Platonic-like realm. Thus, to be a structuralist means that one is committed to understanding the historical context of technology because history is the real product of the structures structuralists deem formative for the world. It is precisely this historical awareness that is lacking in Schumacher and other structuralists, though not in Schuurman. "To be appropriate one must truly know the circumstances in which one works and wishes to correct. Appropriate technology lacks this historical awareness."[68] They too often lack historical understanding because of an implicit but formative anti-intellectualism present in too many parts of the movement. Indeed, sometimes the appropriate technology movement seems like a revival that prefers platitudes to reflection and collegial conversation.

The same problem exists for Schumacher on the sociopolitical level. Our social and political institutions are deeply impacted by technology precisely because technology is so all-pervasive. Modern television has had a profound impact on current political and judicial reality. Appropriate technologists forge no analysis, however, of the impact of technology on socio-political reality.[69] It is as if appropriate technologists dropped down from the theoretical sky a largely correct but decontextualized form of technological redemption to a planet that has yet to be convinced that it has technologically sinned.

Finally, there is a lack of terminological precision in the appropriate technology movement. Words and phrases like "small-scale" and "minimal environmental impact" can contradict each other if precision is lacking. For example, a small aerosol can of spray coolant containing CFCs does *not* have a minimal environmental impact when used. There is nothing inherent in "smallness" that helps or hurts our environment.

The lack of integrality and sufficient structural awareness is surprisingly present in Schuurman's works as well. Schuurman is a trained engineer, philosopher, and part-time member of the Dutch Parliament. He touts an integral ontology. Therefore, we legitimately should expect to see an *integral blend* of structural insights from political/cultural studies, engineering, ethics and philosophy, and science in his alternative section. Practical alternatives and examples should flow from rigorous structural analysis because the whole of reality by its nature must impinge upon the specialized discipline of philosophy.[70] Indeed, there can be no structural analysis (a task Schuurman thinks is important) without a view of integrality present. This reformed Christian tradition equates salvation with wholeness or integrality. It *should* pervade everything. Why then doesn't it pervade Schuurman's alternative sections? Integral alternatives are lacking.[71] Politics, philosophy, engineering, science are not harmonized to bring us the liberation Schuurman rightly calls us to seek. This state of affairs is ironic given Schuurman's technical background, because the very mechanical and industrial overspecialization that he decries is manifest in his work. Interdisciplinary work would help.[72]

Instead, one is left with a valid and necessary but abstract philosophical work on the structural conditions for and thought about modern technology. I believe that Schuurman's view and use of rationality is at fault. It is as if the instrumental rationality with its necessary focused and isolating methodology is the only kind of rationality/methodology that can be used.[73] Schuurman has the experience to offer a holistic interdisciplinary rationality that focuses on *integral* solutions for the matrix of social relationships we call culture.

I have located the problem with Schuurman's work in his view of rationality. A more holistic view of the nature and function of reason is to be found in the following quotation by philosopher Hendrik Hart. He is talking about the relationship between the structures I have outlined and concepts pertaining to these structures:

> Because we have in our culture refined our 'isolation' of these structures so much in the development of theory, we seem tempted to view these structures as well as the concepts and propositions in which we grasp them as themselves 'independent.' We are beginning to appreciate . . . how much both the structures and our understanding of them are related to other realities so thoroughly that we can with justice speak of the relativity of these structures and of our concepts of them to these other realities. In fact, we have come to understand that the more we isolate them and act on them in isolation, the more we endanger ourselves. Only through integration, contextuality, and interrelationship can our rationality function properly.[74]

This lack of alternative integrality leads to at least two ironies. Schuurman is not radical in his alternatives because he is abstract in his theory. Salvation, Schuurman rightly observes, goes to the root or to the heart of life. His philosophical critique indeed does penetrate to the root of philosophical reality but not much beyond. Thus, his solutions do not penetrate to the root of daily concrete problems posed by modern technology because of his lack of integrality.

I tout a view of rationality that is focused and specialized but not without losing its *relational* character.[75] This relational reason complements autonomous and instrumental reason, which forgets that when it makes a mode or an aspect a *gegenstand*[76] of theoretical or specialized thought, one does not escape the matrix of meaning[77] in which and by which specialized thought is accomplished. Schuurman needs to give the same rigor and space to structural and integral alternatives as he does to structural analysis. I am not saying that he is wrong for specializing his thought, or that he has never developed alternatives in isolated papers. Rather, I am saying that in the context of his entire work, alternatives are not integrated with his rigorous analysis.

Furthermore, he does not show the relationship of "evil" to different technologies. He does show how evil on a *grand* scale serves to reduce or truncate reality. He does show how evil reduces particular opportunities in particular areas and technologies. His work in agricultural studies proves this. He does not have an integral view of how evil on the macro level effects "the dislocations" that occur on a relatively small scale. The "direction"—rightness or wrongness—that a particular technology takes or does not take depends on the grand and particular moral circumstances faced by technologists. These circumstances, in turn, interrelate with structural conditions and hence with technical problems.

This reality is especially clear in the development of nuclear power. A deep and persistent arrogant grand attitude about human relationship to the natural environment is related to viewing the environment as a means to ever-increasing economic ends. This grand economic reductionism has led to a myopic focus on gaining more fuel at all costs. This reality has led to an obsession with fossil and nuclear fuels. Consequently, solar sources have been closed down or not sufficiently developed. Had Schuurman linked a detailed analysis of the given and the distorted conditions for technology with practical alternatives, then he would have made a considerable contribution to the discipline.

Finally, I am not sure that Schuurman gives an accurate, current account of all the norms or principles associated with specific aspects. For example, it is held by specialists within his school of thought that the term "harmony" when applied to the aesthetic dimension of life is overly Greek and rationalistic. Perhaps the term *allusiveness* better depicts the core reality of aesthetic life.[78] I am not being overly scrupulous in our objections at this point because the exact nature of our response to technology depends on the meaning of the norms inherent in other areas of life.

There can be no doubt that Schuurman has added to our understanding of the meaning and place of modern technology. I can only hope that a more holistic reflection will be manifest in future works.[79]

5
Conclusion

I have surveyed four dominant positions in technological interpretation. These views speak to the effect of technology on our personal, cultural, and historic lives. These descriptive types have not been "ideal," in that they have been removed from practical reality. Quite the contrary is the case. Real values, principles, and assumptions have been linked with practical policies that have been decisive for our lives. In light of this analysis, I have asked what, in our opinion, the central question of technology should be for the twenty-first century: what place ought technology to occupy in our lives? This ethical question is forced on us by the imposition and the character of more than twenty centuries of technology. The question must be addressed ethically and publically. It can be answered only when technology is placed in context with other areas of life and together we develop a vision for the place of technology. If we decide to address technology in this manner, and I believe we must, then the relative values of other areas of life must and will color our considerations.

All positions have strengths and weaknesses. Optimists argue well when they say that technology discloses new possibilities for life. Machinery can enhance, for example, the work of humanity. However, Julian Simon goes too far when he argues that expanding technological options will correct if not eradicate traditional human problems. It is myopic to believe that technical solutions to nontechnological problems represent sufficient answers. It is myopic because reality is more diverse than optimists understand. Indeed, Simon mistakenly equates total human betterment with economic and technical improvements. This equation does not signal *progress*; it signals social and human regress. The simplistic identification of human, natural, and cultural identity with technically related processes creates a false hope and reduces the dignity of all involved. Humans are more than *Homo*

faber—man the fabricator—and the world around us more than a
mere machine, or fodder for the technological machine. I have called
this myopia *reductionism* or the reduction of complex reality to technical meanings. Thus, structurally diverse problems require a more
multifaceted, interdisciplinary solution.

Optimists also underestimate the destructive effects of technology.
This deadly underestimation downplays, for example, the many costs
needed in the clean up of radioactive wastes. Once touted as "the
atoms for peace" and "too cheap to meter," cleaning up after the results of the splitting of the atom will cost us over one and one half
trillion dollars in the next seventy years. This figure does not even
begin to calculate the damage *already* caused by leaking radioactive
wastes into ground and water sources.

Optimistic assumptions lead to an ever-increasing amount of social
technique and technical hardware that dominate our globe. This exaggeration is welcomed by optimists but produces a dramatic irony
for the pessimist. The gargantuan onslaught we call modern technology is a gulag—a prison—argues noted pessimist Jacques Ellul.
Modern technology robs us of our freedom and identity, though its
birth in Western society came with the promise of freedom. Congested
traffic caused by our chariots of mobility—automobiles—is one good
example of this irony, argues the pessimist. In exchange for our
human freedom we have spawned a technological society. Our society
lives by, for, and unto technology. This domination is called *idolatry*
by Ellul and repressive by many postmodernists.

I believe the pessimist's critique must be heard because the
overextension of technology is indeed a problem. An example of this
problem is the dominance by experts with the consequent loss of
democratic responsibility. Unquestioned technology, like alcoholism,
can insidiously endanger our lives. Many people wisely question the
amount and the place alcohol occupies in our lives; why can we not
question with equal fervor the amount of and the place taken by
technology in our lives? Perhaps the answer is that we do not realize
the potential of modern technology. The word "Chernobyl," in this
regard, has significance. The word translated means "wormwood."
In **apocalyptic** literature (i.e., Revelation 8:10–11), wormwood is
the evil that falls from the sky to poison the rivers and the waters,
the result of which was death by a third of the "earth dwellers."
Indeed, the *growing* incidence of at least thyroid cancer among
Chernobyl and Kiev children caused by the fallout resulting from
the 1986 power-plant explosion forces us to label modern technol-

ogy apocalyptic. The fabric of our lives is too often exchanged for the narcotic and destructive effects of technique!

However, pessimism can only leave us with dialecticism and ambivalence. When one absolutizes evil, as Ellul does, then one does not have "one all-encompassing" world view (to use Ellul's phrase). Rather, one is left with conflict, anxiety, and a lack of wholeness. Similarly, the postmodernists offer us primarily ambivalence toward modern technology. They see little that is positive in modern technology beyond a reflection of their autonomous freedom. I have argued on the contrary that *all* of reality is inherently positive, though we experience the good in technology in a deeply distorted manner. Evil seems more a fateful distortion than a sovereign master. We cannot be guided by the disoriented state of postmodernism.

Realism offers us a seemingly more balanced approach to technological assessment. It readily admits that technology has good and bad effects. It asks us to consider what we are willing to give up or trade off to reduce our risk of technological harm. Conversely, it asks us how much we are willing to part with to secure the positive fruits of technology. It develops a multidisciplinary methodology to assess risks and prevent harm. But it lacks a firm democratic footing because there are no social pluralities, or diverse communities present during debate.

The multidisciplinary effort of risk assessment offers much for our use and understanding, especially in an age of overspecialization. We are forced to decide collectively which technology we want for what price and then select the relevant principles that will guide our deliberations. Further, tradeoffs force us to consider the place that technology should occupy in our priorities. Trading off implies giving up a little bit of X so that more of Y can be had. Implied in this methodology could be a multifaceted view of reality, one rich in principles for life and richer in public choice.

However, it is precisely this rich texture for life that is abandoned in realism's reductionistic and utilitarian view of the person. This problem affects, consequently, how tradeoffs are made. This form of utilitarianism views the person as a utility-maximizing, pleasure-seeking, pain-avoiding, economically calculating, rationalistic individual who weighs technical utilities against extratechnical disutilities. Tradeoffs, safety, and rewards are calculated in dollars and cents with decisions made accordingly. This view has more in common with an eighteenth-century Enlightenment view than with modern understanding. The humanities and social and natural sciences have given us too much insight to view the rich complexity of humanity as a rational, utility-maximizing individual: a quintessential reductionistic view. That our more rich, accurate, and meaningful view of reality will be difficult to convert into an evaluative methodology is conceded. However, a much more rewarding project than trying to cram humanity into a tight utility-maximizing box would be to develop a more diverse view of reality and humanity such that technology discloses a more multi-faceted view of reality. This fuller view would produce a more rewarding and accurate assessment methodology, one in which ethical principles inherent in reality guide decision making.

Further, realism sets great faith in democratic debate. At the same time realists want to avoid "dogmatic ideologies." Some realists still believe that positions can be taken that are free of values and bias. I find this positivistic view misleading if not naive. Congressional hearings on proposed new technologies are shaped often by communities or collective groups of people who have an organized, more-or-less coherent identity and mission. This pluralistic political reality is not to be condemned, as too many realists do. Rather, it is to be welcomed because pluralistic community-centered debate represents the nature of reality. It is good and necessary that environmentalists and engineers (for example) square off over the implications of a new technology. We must have dialogue about the place and meaning of technology, but there is no "salvation" in dialogue per se, at least not sufficient healing. Wholeness will only be increased as communities bring their rich, principled views of life to the discussion table with an eye to how their necessarily particular view of life can benefit public life. This must happen if we are to evolve beyond our current state of being collective technical sheep.

Further, realism's view of reality as inherently problematic troubles us. It is not reality that is inherently problematic. Human mistakes, limitations, and poor judgement give us trouble. We do not need a method to control reality. Rather, we need an adjusted view of technology to reduce the problem—the unnormative, overextended view of modern technology. It is our autonomy not technology or reality that is at fault.

I believe that structuralists offer the greatest hope for a reform of technological use. Schumacher and Schuurman can affirm all that is good in the previously stated positions while avoiding the pitfalls. For example, they can make use of the inherent goodness of technology while not falling prey to technological optimism. At the same time, they speak eloquently about the "idolatry" and the "mega nature" of modern technology while offering a vision of how technology can be used to enhance life, not dominate it. They would use realism's multi-disciplinary methodology, without slipping into realism myopia. Their vision is focused on a view of reality that is structured by God. Technology is located in a place or a "room" because reality is simultaneously meaningful yet delimited. Living leads to a discovery of principles for life, or ethical principles for living. God is said to disclose principles for how technology may be used. These principles command our lives and thus provide the context for freedom. It is deeply ironic, yet profoundly true, that to be free one must listen and obey, argues the structuralist. Because principles such as stewardship, humility, and justice are the conditions for our existence, ethics is *not* an exercise that is appended to an otherwise value-neutral technical process. Rather, principles provide the condition for our existence and thus form the conditions for our lives.

Because God is sovereign, evil cannot be. Structuralists, therefore, resist technical determinism, à la Ellul. Indeed, God's provident involvement in the maintenance and care for reality gives structuralists a proper sense of tamed optimism in that reality is always something that elicits surprise and joy.

Furthermore, questions focusing on locating technology within life lead us deeper into ethical inquiry. Principles give meaning and perspective to life. The study of ethics is meaningful because the principles for the good life constitute meaningful reality. Structuralists believe they have found and are developing technology with principles like stewardship and historical sensitivity to culture and people. Thus, structuralists overturn one glaring problem and strengthen one ability of realism. Structuralists correct realism's pragmatism by providing a firm and constitutive ethical foundation. In turn, this foundation gives fuller vigor to realism's attempt at an interdisciplinary framework. The phrases "normative demands" (Schuurman) and "four levels of being" (Schumacher) both describe our composite ethical existence and serve to restrict and give positive meaning and purpose to life. Ignoring these norms is *the* problem for technology; the existence of technology per se is not the problem.

I have noted also that structuralists' alternatives were less well developed, their critique exceeds their options. The "small-is-beautiful" movement, while beginning to offer alternatives especially for Third World countries, suffers from marked terminological and methodological problems. Sufficient philosophical and scientific rigor could remedy this situation, but this rigor is not to be found, especially since Schumacher's death. Similarly, the engineering, philosophical, and parliamentary experience and training of Schuurman should evidence more practical social and engineering alternatives. I have located this problem, ironically, in the lack of sufficient structural alternatives.[1]

Our attention in this work has been focused primarily on the *foundational* issues apparent in the representative types. I have not attempted or even approached a systematic alternative view of technology. We have begun, however, to set forth some intellectual and practical alternatives. This work has dealt with foundational or basic religious and philosophical ideas. Technology must enhance life, not dominate it. To enhance life technology must: be affordable (unlike the nuclear industry); preserve and enhance aesthetic beauty (unlike modern home-building subdivisions); promote rather than divide family life by developing not destroying family commitment (like the current use of TV); be enriching not depleting of the natural environment, and promote a more ethically rich, compassionate, less coldly mechanistic view of the universe.

We will have to increase the pace of interdisciplinary cooperation if life is to be richer. That is, we must bring the necessary disciplinary specialization to a round table and thereby escape our disciplinary ghettos of isolation, our well-protected turfs of academic autonomy.

To become more intellectually cosmopolitan will require two additional steps with accompanying resources. Our graduate programs will have to be redesigned to make connections between disciplines, as much as they concentrate in unearthing new bits of specialized knowledge. To do this will require emotional changes as well. We can become defensive, protective, and unnecessarily critical when we feel someone is trespassing on our intellectual turf. New interdisciplinary insight often is treated as substandard because it is allegedly not as accurate as specialized knowledge. Specialization without connections leads us to a fragmented world. Second—and this will be difficult in our economy—we need more administrative dollars for positions *within* and *without* the department structure to promote interdisciplinary cooperation. Connections are thwarted when one is located in and confined by the department structure.[2]

These positions must be created, because reality is, at its very core, coherent. Though our necessary specialties carve apart reality, a view of its coherence must be regained. This is said because a fuller experience of meaning, purpose, relevancy, and quality are at stake. If structuralists are correct, then a rich world of meaning, purpose, relevancy, and quality can*not* be created; it is to be discovered! In our push for professional specialization we have thought that quality and high standards arise from specialization. That this thought originates from the Industrial Revolution should not be lost on us. Years of teaching, philosophical conviction, as well as colleagues' evolving insight, and most current educational theory convince me otherwise. When students of all ages have experienced the coherence of education—how education fits together—their expanded visions of themselves and their work rockets them into more meaningful horizons, with accompanying academic skills, than previously imagined. This can be said of graduate, undergraduate, and continuing education students. This fact gives me hope enough to believe that someday we might be wise enough to change our namesake from Prometheus to that of Solomon: wise rulers who can rightly discern life.

Notes

INTRODUCTION

1. Professor Jack Levin, "Unabomber A Hero to Some," in *USA Today*, April 11, 1996, 1.

2. Egbert Schuurman, *Technology and the Future: A Philosophical Challenge*, trans. Herbert Donald Morton (Toronto: Wedge, 1980).

3. Ian Barbour, *Ethics in An Age of Technology* (San Francisco: Harper & Row, 1994).

4. Ontology is the study of being. Specifically, in this study we are talking about the context for technology. The context is comprised of aspects or ways that all life functions. The conditions, meanings, and limits to functioning are important concepts for this study of ontology. Students often like to say that all of reality is like a home while different aspects are like rooms within that home.

5. Yaron Ezrahi, Everett Mendelsohn, and Howard Segal, *Technology, Pessimism, and Postmodernism* (Hingham, MA: Kluwer, 1994) is an example of such a piece that, while valuable for this study, does not pretend to go beyond the study of pessimism.

6. For example, students seem to master and enjoy Jacques Ellul's thought while the work of Martin Heidegger remains a bit of a mystery to them (and to me and some others of the guild, I suspect).

7. At the moment of the explosion of the *Challenger* shuttle, I was driving to deliver a lecture to a class of engineers on the subject of technological optimism! This irony was not lost on me or on them.

8. Stephen H. Cutcliffe, et al., eds., *New Worlds, New Technologies, New Issues* (Bethlehem, PA: Lehigh University Press, 1992).

9. This view is developed more completely in my *Between God and Gold: Protestant Evangelicalism and the Industrial Revolution*, foreword by Martin Marty (Madison, NJ: Fairleigh Dickinson University Presses, 1993).

The work of Michael Polanyi, *Personal Knowledge* (Chicago: University of Chicago Press, 1958), 65, states: "It is the act of commitment in its full structure that saves a personal knowledge from being merely subjective. Intellectual commitment is a responsible decision, in submission to the compelling claims of what in good conscience I conceive to be true."

10. The work of Gerard Radnitzky, "Towards a Praxeological Theory of Research," in *Systematics*, 10 (1972), 131, states: "The complex of hypothesis, problems and instruments resembles the sketch-map on which the prospector has marked sites he thinks are worth exploring. The sketch-map is based upon the researcher's world-picture, hypothesis and ideals for science associated with them."

11. See James MacLachlan, *Children of Prometheus: A History of Science and Technology* (Toronto: Wall and Emerson, 1989) and the classic work from David S. Landes, *The Unbound Prometheus: Technological Change and Industrial Development in Western Europe from 1750 to the Present* (London: Cambridge University Press, 1969). It was the latter book, first surveyed in graduate school, which led me to see the nature of optimism and its trust in human rationality.

Chapter i. Technological Optimism

1. Are animals and other less biologically developed forms of life capable of technical activity? This question and the debate that surrounds it is an instructive one. For an excellent, succinct discussion of relevant issues, see Frederick Ferre, *Philosophy of Technology* (Englewood Cliffs, NJ: Prentice-Hall, 1988), 17–18, 31, 85.

2. See chapter 1 of my *Between God and Gold: Protestant Evangelicalism and the Industrial Revolution, 1820–1914* (Madison, NJ: Fairleigh Dickinson University Press, 1994).

3. For a provocative discussion of the origins and growth of technical reason, see Friedrich Klemm, *A History of Western Technology*, trans. Dorothy Waley Singer (Cambridge: MIT Press, 1964), 231–66.

4. Idealism is a philosophical movement that stresses the centrality of the mind or the spirit for all of reality. Thus, reality is essentially mental in nature. If humans and other reality are essentially mental or rational, then we can use our reason to penetrate or understand reality, so the theory goes. Further, there are other forms of idealism. Among them are the dualistic idealism of Descartes, the realistic idealism of John Locke, and subjective idealism of David Hume.

5. See Klemm, *History*, 231–34, in the "Introduction to the Age of Reason" for a deepening of this analysis.

6. Lynn White's charge that Christian thinking is responsible for the environmental crisis is only partially correct. The Enlightenment must also be studied to discern problematic patterns of environmental relationship. See Bob Goudzwaard, *Capitalism and Progress*, trans. and ed. Tosina Van Nuis Zylstra (Grand Rapids, MI: William B. Eerdmans, 1979), 36–54. For the view of key scientists and their view of nature see Loren Wilkinson, ed., *Earth Keeping: Christian Stewardship of Natural Resources* (Grand Rapids, MI: William B. Eerdmans), 124–34.

7. For a review of additional late medieval and early modern technical improvements that helped promote autonomy, see Klemm's discussion of horse-collar, heavy wheeled plough, iron horseshoes, rotation agriculture, the lathe, improvements in the vault, gunpowder, stone masonry, and, perhaps most important of all, the printing press. In Klemm, *History*, 79–117.

8. René Descartes, *Discourse on Method and the Meditations*, trans. F. E. Sutcliffe (Harmondsworth, U.K.: Penguin, 1979), 78.

9. For a penetrating discussion of the contours of the mechanical philosophy, see Margaret C. Jacob, *The Cultural Meaning of the Scientific Revolution* (New York: Knopf, 1988), 52–54 and 232–34. See also Lewis Mumford, *Technics and Civilization* (New York: Harcourt and Brace, 1934), 46–48 and passim.

10. This doxology to technology is given in engineer: Samuel C. Floreman's "In Praise of Technology," in *Technology and Change*, ed. John G. Burke and Marshall C. Eakin (San Francisco: Boyd and Fraser, 1979), 21. It must be noted in fairness to Floreman that he is not an optimist—at least that is what his more recent works suggest—but it is also worth noting that the cartoon that introduces this article evidences the same kind of linear view of progress discussed here.

11. Clarence E. Ayers, "The Industrial Way of Life," in *Technology and Change*, ed. Burke and Eakin, 425f. Emphasis added.

12. Hugo A. Meier, "Technology and Democracy, 1800–1860," in *Technology and Change*, ed. Burke and Eakin, 212.

13. See Goudzwaard, *Capitalism and Progress*, xxii, 36, 151f, and 161 for the religious grounding for the notion of economic and technological progress.

14. There are several books that are helpful in discerning the contours and development of this notion of progress. The classic study has been written by J. B. Bury, *The Idea of Progress* (London: Macmillian, 1920). For secularization of the notion of ethical imperative of progress, see Goudzwaard, *Capitalism and Progress*. For a contemporary interdisciplinary analysis of the notion of progress, see David H. Hopper, *Technology, Theology, and the Idea of Progress* (Louisville, KY: John Knox Press, 1991).

15. See Klemm, *History*, 135–50, for a well-documented discussion of Renaissance technology.

16. Valeri Legasov, Leo Feoklistov, and Igor Kusmin, "Nuclear Power Engineering and International Security," in *Soviet Life* 353, no. 2 (February, 1986): 14.

17. I enjoyed a recent conversation with Professor Goudzwaard about Soviet technology in general and the Chernobyl disaster in particular. During the conversation he spoke of a trip he took to the former Soviet Union. He was invited to the Chernobyl site, where he conversed with the nuclear experts about the aftermath of the disaster. The Soviet official admitted in a moment of honesty, "perhaps we have been too hasty and too arrogant in our assumption about technology's safety." Is optimism born out of a false pride? Does optimism lead to disillusionment? These are some of the important questions that must occupy our dialogues.

18. Hooper, *Technology, Theology*, 24, as quoted in the *New York Times*, June 11, 1986, B6:1. Emphasis in original.

19. The phrase is taken from the book by the same name. See Julian Simon, *The Ultimate Resource* (Princeton: Princeton University Press, 1981).

20. These arguments can be found in Simon, *Ultimate Resource*, 3–5.

21. Simon, *Ultimate Resource*, 9.

22. Simon, *Ultimate Resource*, 23. Emphasis in original.

23. Simon, *Ultimate Resource*, 38, quoting Herman Kahn, William Brown, Leon Martel, et al., *The Next Two Hundred Years: A Scenario for America and the World* (New York: Morrow, 1976), 101.

24. Quote from Lewis Mumford, *Technics and Civilization* (New York: Harcourt Brace Jovanovich, 1963), 134–37. For an optimistic view of the effect of the printing press on society starting in the Middle Ages, see Klemm, *History*, 97–98.

25. See Klemm, *History*, 175–80 for Galileo's relationship to the rational mathematical-mechanical world view, one necessary for the advent of the computer of Descartes.

26. For a fascinating study linking Taylor's rather rigid personality to his mechanistic views on labor, see Sudhir Kakar, *Frederick Taylor: A Study in Personality Innovation* (Cambridge: Harvard University Press, 1970).

27. Arnold Pacey, *The Culture of Technology* (Cambridge: MIT Press, 1984), 15. He quotes the authoritative works of W. G. Hoskins, "Harvest Fluctuations and English Economic History," *Agriculture Review* 16 (1968): 15–45; and Susan Fairlie, "The Corn Laws and British Wheat Production," *Economic History Review*, ser. 2, 22, (1969): 109–16. For further information see Pacey, *Culture of Technology*, 181.

28. Ferre, *Philosophy of Technology*, 57.

29. See Karl Marx, *Das Kapital* (London: Eden and Cedar Paul, 1930), 177.

30. Ferre, *Philosophy of Technology*, 60, quoting R. Buckminster Fuller, *No More Secondhand God and Other Writings* (Garden City, NY: Doubleday, 1963), x.

31. Ferre, *Philosophy of Technology*, 61, quoting Fuller, *No More Secondhand God*, vii.

32. The reader may perceive that Reason is treated as an instantiated person or force. That is, it seems as though Reason is not some abstract philosophical principle but has become a semipersonal deity that concretely and practically orders and therefore cares and provides for humans. Thus, does it not seem as though a deistic form of providence is at work? If the reader perceives this claim of reason to be in effect, then he/she is not confused but perceives the claim of the optimist accurately.

33. Ferre, *Philosophy*, 60, quoting Fuller, *No More Secondhand God*, 17. The more advanced, perceptive reader may want more study in technological optimism. Egbert Schuurman's *Technology and the Future* (Toronto: Wedge, 1980), 177–312, and his discussion of "The Positivists" should be consulted. My difference with particular points of Schuurman's excellent work will be seen when I speak of the structuralists in chapter 4.

34. Howard P. Segal, "The Cultural Contradictions of High Tech: Or The Many Ironies of Contemporary Technological Optimism," *Technology, Pessimism, and Postmodernism*, ed. Yaron Ezrahi, Everett Mendelsohn, and Howard Segel (Boston: Kluwer, 1993), 175–214.

35. The term *ontology* formally means the study of being. Less abstractly, ontology is the study of the context and the conditions for life and existence.

36. I use the rational/irrational distinction because it represents the conventional manner of speaking about order and disorder or predictability/unpredictability. I do not believe, however, that this dialectic represents the best way to think about the problems at hand.

37. Pacey, *Technology*, 14f, also has an excellent critique of this notion of linear progress.

38. Herman Dooyeweerd, *Roots of Western Culture: Pagan, Secular, and Christian Options* (Toronto: Wedge, 1979), 150. My thanks must be extended to Egbert Schuurman for allowing me to review the advance copy of his paper, "The Technological Culture Between the Times: A Christian Philosophical Assessment of Contemporary Society," which analyzes the problems related to autonomy dealt with in this section of my book.

39. Egbert Schuurman, "The Technological Culture Between the Times: A Christian Philosophical Assessment of Contemporary Society," 6.

40. Schuurman, in a debate with Bob Goudzwaard of the Free University in the Netherlands, maintains that technicism is the grand exaggeration or idolatry of Western civilization. I cannot agree with this assessment. Although technicism certainly represents a powerful force and is ontologically prior to economic actions, there are at least two reasons for believing economism is the dominant exaggeration in our lives. First, most people practically spend much more time, effort, and anxiety in pursuit of wealth than they do technology (although the two cannot be separated in reality). This point will become more forceful in the chapter on realism, when I will show that even modern technology succumbs to acquisitive rationality. Second, and more importantly, precisely because instrumental rationality is a means to a greater end, it is not an end in life. Means are secondary, ends are primary. Thus, Schuurman's own "structural" analysis shows that technology is simply a means to some greater end.

41. See Michael Bookchin, *Toward an Ecological Society* (Montreal: Black Rose, 1980) and his *The Ecology of Freedom: The Emergence and Dissolution of Hierarchy* (Palo Alto, CA: Cheshire, 1982). It has been my privilege to teach interdisciplinary courses in technology for ten years. The predominant student desire is to want to return to some form of relatively primitive naturalism. This desire occurs because of the problem of technological subjugation of nature. Most want to consider the Amish way of life or want to return to the totally organic farm, hopefully placed far away from cities and their suburbs.

A faculty member of Carnegie-Mellon University, one of the most prestigious technical universities in the United States, approached me while I was working on the first draft of this manuscript. When told of what I was doing, he offered a further insight on naturalism. Several recent publications in the philosophy of technology suggest that the human brain is *determined* just the way a computer decoder is determined by bombarding signals. Thus, freedom is an illusion because both nature and culture determine the brain's responses through constant sensual stimuli. See, therefore, Sherry Turkle, *The Second Self: Computers and the Human Spirit* (New York: Simon and Schuster, 1984), 285–92.

CHAPTER 2. TECHNOLOGICAL PESSIMISM

1. Jacques Ellul, *The Technological Society*, trans. John Wilkinson (New York: Vintage, 1964), xxii. Emphasis in original.

2. The reader should note that I am presenting a brief summary of the positions I will explore in this book: facile optimism for technological optimism; calculating the trade-offs between the positives and the negatives for the technological realists; and placing technique within life's constraints for structuralists.

3. Ellul, *Technological Society*, 14.

4. Ellul, *Technological Society*, 143. While I appreciate Ellul's critique of the idolatry of technology, there seems to be a contradiction in his view: sacred/secular, spiritual/temporal. Ellul's point, as we will soon see, is that technique has become a monism or one universal all-encompassing force. Therefore, an equally forceful, contradictory principle must arise to dialectically oppose technique.

5. *Ideology* is that body of doctrine, myth, or symbol of a social movement that has been put into practice by some cultural or political plan. For Ellul's treatment of

technology as ideology, see his *The Technological Bluff*, trans. by Geoffrey W. Bromiley (Grand Rapids, MI: William B. Eerdmans, 1990), 172–88.

6. See Kai Nielsen, "Technology as Ideology," in *Research in Philosophy and Technology*, (Greenwich, CT: JAI Press, 1978), vol. 1, 131–47. It is interesting to note, in this regard, how both Joseph Stalin and Henry Ford made use of Frederick Taylor's time and motion studies to boost production. See Klemm, *History*, 325, 333–35.

7. Jacques Ellul, *Technological Bluff*, trans. Geoffrey W. Bromiley (Grand Rapids, MI: William B. Eerdmans, 1990), 257f. Ellul, of course, views the modern consumption mania as the consequence of the saturation of the market with "productive technologies."

8. Ellul's theological beliefs are much more complex than this segment on ideology suggests. For the moment it must be stated that from the side of history, God's absence from human history is occasioned by technology.

9. Ellul, *Technological Bluff*, 257f.

10. Ellul, *Technological Society*, 307.

11. See, for example, Richard S. Barnett and Ronald E. Muller, *Global Reach: The Power of Multinational Corporations* (New York: Simon and Schuster, 1974), 162–72.

12. Ellul, *Technological Society*, 134. See also Langdon Winner, *Autonomous Technology: Technics-Out-of-Control as a Theme in Political Thought* (Cambridge: MIT Press, 1977), 16–17.

13. Ellul, *Technological Society*, 56, 90.

14. Ellul, *Technological Bluff*, 371–72.

15. Ellul, *Technological Society*, 93–94.

16. I received a "survey call" from a nationally known cable company the day this chapter was being drafted. A representative wanted to know my opinion on the value of cable TV. The person questioned me about everything from my salary level to my religious (used here in the traditional sense) convictions. While I was answering questions, my wife told me to hang up because "you are being used." Responding I whispered, "A statement is being made. Leave me alone." The discussion finally had to be shortened because another commitment awaited. I promised to be near the phone on Sunday, no less, when he would call again. Total elapsed time for this abbreviated survey was one hour and five minutes. Was I, or was my wife, correct? Where do you suppose this valuable information went?

17. Norbert Wiener, *Cybernetics, or Control and Communication in the Animal and the Machine* (Cambridge: MIT Press, 1950), 22, 55, and passim.

18. D. J. Wennemann, "An Interpretation of Jacques Ellul's Dialectical Method," in *Broad and Narrow Interpretations of the Philosophy of Technology*, vol. 7, ed. Paul T. Durbin (Boston: Kluwer, 1990), 181f.

19. Ellul would not allow his ways of thinking to be called methods because this would involve him in a technique of thinking. See Wennemann, "An Interpretation," 181–83. It is unfortunate, therefore, that Wennemann has chosen to describe Ellul's thought as involving a method.

20. Wennemann, "An Interpretation," 182, quoting Ellul, "Mirror of These Ten Years," in *Christian Century*, February 18, 1970, 200.

21. Wennemann, "An Interpretation," 183, quoting William H. Vanderburg, ed., *Perspectives on Our Age: Jacques Ellul Speaks on His Life and Works* (New York: Seabury, 1981), 15. Wennemann's entire note should be used to clarify Ellul's phrase "two totalities."

22. Wennemann, "An Interpretation," 185.

23. See Clifford G. Christians and Jay M. van Hook, eds., *Jacques Ellul: Interpretive Essays* (Champaign: University of Illinois Press, 1981), 296.

24. Jacques Ellul, *What I Believe*, trans. Geoffrey W. Bromiley (Grand Rapids, MI: William B. Eerdmans, 1989), 4.

25. Ellul, *What I Believe*, 3.

26. Ellul, *What I Believe*, 4.

27. Ellul, *What I Believe*, 24.

28. Ellul, *What I Believe*, 28. Ellul derives his dialectical theology from the modern prolific reformed theologian Karl Barth.

29. Ellul, *What I Believe*, 33.

30. Ellul, *What I Believe*, 45.

31. Ellul, *What I Believe*, 49.

32. Jacques Ellul, *The Ethics of Freedom*, trans. and ed. Geoffrey W. Bromiley (Grand Rapids, MI: William B. Eerdmans, 1976) [French original, 1973]), 12.

33. Because God is transcendent, She transcends our pronouns. Furthermore, since He includes all people in His care, men and women, girls and boys are to be equally valued and respected. Hence, the pronoun choice is somewhat arbitrary for me.

34. Ellul, *Ethics of Freedom*, 14. Emphasis added.

35. Ellul, *Ethics of Freedom*, 16.

36. Ellul, *Ethics of Freedom*, 16. While traveling in the Netherlands to study and to lecture recently, I enjoyed an evening with Protestant reformed philosopher of technology Professor Egbert Schuurman. He told me that (the late) Professor Ellul was then beginning to admit that the "New Jerusalem" may be more evident in history than he has been able to see. In fact, there may be a book from the pen of Ellul on this topic appearing in the not too distant future. Perhaps the Christian half of Ellul is beginning to gain ascendancy in his twilight years?

37. Ellul, *Ethics of Freedom*, 34–35 and passim.

38. Ellul, *Ethics of Freedom*, 75.

39. Ellul, *Technological Bluff*, xiii.

40. For a thorough, well-documented book on how the Industrial Revolution overran "vulnerable" and "weak" educators thereby taking control of public education, read Raymond E. Callahan, *Education and the Cult of Efficiency* (Chicago: University of Chicago Press, 1962).

41. Ellul, *Technological Bluff*, 15.

42. For a mainline economist's critique of this optimistic notion of perfect competition and the reality of imperialism read, Douglas C. North, *Institutions, Institutional Change and Economic Performance* (New York: Cambridge University Press, 1990), 131ff, esp. 134–35.

43. Ellul, *Technological Bluff*, 21.

44. There are many writers who are pessimistic about technology's meaning for our lives. The number of pessimists seems to grow in the latter half of the twentieth century. My choices in this chapter are only meant to be introductory and illustrative. The experienced reader will want to consult a more thorough listing. See, therefore, Egbert Schuurman's notion of the "transcendentalists" or thinkers who see technology as a threat to human freedom in *Technology and the Future: A Philosophical Challenge*, trans. Herbert D. Morton (Toronto: Wedge, 1980), 51–176.

45. Jürgen Habermas, *Technology and Science as "Ideology"* (Frankfurt: Suhrkamp Verlag, 3rd ed., 1969), 7.

46. Habermas, *Technology*, 3.

47. Habermas, *Technology*, 53.

48. Habermas, *Technology*, 80 and passim.

49. Habermas, *Technology*, 81 and 96.

50. Habermas, *Technology*, 91.

51. Habermas, *Technology*, 101–2.

52. Nicholas Berdyaev, "Man and Machine," in *The Bourgeois Mind and Other Essays* (New York: Sheed and Ward, 1934), 203–4.

53. I am not claiming that this analysis of the thought of Habermas is exhaustive. Given my limits, I can only note that one school, the "Frankfort School" of neo-Marxist thought, changed its perceptions about the impact of modern technology for a time. This change needs to be noted as the philosopher charts the evolution of Marxism.

54. Berdyaev, "Man and Machine," 204.

55. Berdyaev, "Man and Machine," 205.

56. Berdyaev, "Man and Machine," 210.

57. Berdyaev, "Man and Machine," 212.

58. Berdyaev, "Man and Machine," 212.

59. Perhaps the clarion call of postmodernity was sounded by Richard Rorty in *Consequences of Pragmatism* (Minneapolis: University of Minneapolis Press, 1982).

60. Yaron Ezrahi, Everett Mendelson, and Howard Segal, eds., *Technology, Pessimism and Postmodernism*, (Boston: Kluwer, 1994), 2.

61. Perhaps the best anti-Enlightenment piece on postmodernity is David Harvey, *The Condition of Postmodernity* (Oxford: Basil Blackwell, 1980).

62. This point is outlined in Segal, "Introduction," in *Technology, Pessimism, and Postmodernism*,1–10; and in Leo Marx, "The Idea of 'Technology' and Postmodern Pessimism" in the same volume, 11–28.

63. Leo Marx, "Technology," 21. Emphasis added.

64. His book *Technics and Civilization*, published in 1934, set the tone for modern pessimism.

65. The word *Pentagon* is being used here as a pun. The word stands both for the literal military establishment called the Pentagon and modern technology's mighty and destructive five P's: power, profit, productivity, political control, and publicity. See Everett Mendelson, "The Politics of Pessimism: Science and Technology Circa 1968," in *Postmodernism*, 151–52.

66. See Jonathan Allen, ed., *March 4: Scientists, Students and Society* (Cambridge: MIT Press, 1970), "Union of Concerned Scientists, Faculty Statement," xxii–xxiii.

67. Herbert Marcuse, *One Dimensional Man* (Boston: Beacon, 1964), 165–86.

68. Herbert Marcuse, *An Essay On Liberation* (Boston: Beacon, 1969), 12. It is necessary to state that Marcuse's intellectual roots can be traced to the Frankfurt school of thought. This neo-Marxist school manifests a great deal of ambiguity about modern technology. I will take up this ambiguity about the effect and the place within our lives of modern technology in my concluding section. For now I must note that treating Marcuse under the topic of pessimism leaves me with some reservation.

69. Everett Mendelson, *Postmodernism*, 155.

70. Unfortunately I have not always recognized that women were accorded the same status as that of men—that being the image of God.

71. Berdyaev, "Man and Machine," 206.

72. The television series *Miami Vice* aired in America from 1984 to 1989. Set in the steamy confines of Miami, Florida, two vice cops—Crocket and Tubbs—fought the sale and distribution of a myriad of illegalities. What set these cops apart from most mortals I know was their exposure to technologically titillating lifestyles. Fast cars, expensive boats, gorgeous homes, mouth-watering food, and lavish sex (itself methodically choreographed) provided a technological milieu that many aspirants would find attractive. In comparison with this attraction, the Zambian moralizing about the evils of capitalism seems quixotical.

73. Descartes' thought was covered in Chapter One. His thought gave rise to the modern mechanistic worldview. Human freedom was sacrificed in this view to natural mechanical necessity.

74. Ellul's reformed Protestant tradition is close to that of my own. It is for that reason that I both respect his work and at the same time find it so troublesome.

75. Segal, "Introduction" to Ezrahi, Mendelson, and Segal, eds., *Technology, Pessimism and Postmodernism*, 1–10, especially 9–10.

76. Ellul, *What I Believe*, 1.

Chapter 3. Technological Realism

1. Raphael G. Kasper, *Technology Assessment: Understanding the Social Consequences of Technical Applications* (New York: Praeger, 1972), 3–4. Emphasis in original.

Throughout this chapter mention will be made of "the realist" as if that label represented some specific person. This is not the case. Spokespersons for realism, unlike optimism and pessimism, tend to be relatively more ubiquitous and hence less noteworthy. Relative to optimists and pessimists, the names of realists tend to be less well known. They are less easily identifiable and intellectually traceable. Hence the generic term "the realist" occurs frequently.

2. It should become increasingly clear that there is a difference between technological assessment and risk analysis. The former signifies the general assessment of technology's impact upon society while the latter involves the specific analysis of the probability of exposure to harm or risk.

3. A readable glossary of assessment terms can be found in United States General Accounting Office publication *Health Risk Analysis: Technical Adequacy in Three Selected Cases* (Washington, DC: General Accounting Office, 1987), 166–71.

4. See Ron Westrum, *Technology and Society: The Shaping of People and Things*, (Belmont, CA: Wadsworth, 1991), 326–27.

5. For a more comprehensive treatment of these points, see Ann Neale, *Technological Assessment: Some Political and Theological Reflections* (Richmond, VA: Center for Theology and Public Policy, 1980), 1–3.

6. For a introductory exposition to the techniques and goals of technological assessment, see Westrum, *Technology and Society*, 328–29.

7. These functions are summarized in Kasper, *Technology Assessment*, 9–19.

8. See H. W. Lewis, *Technological Risk* (New York: Norton, 1990), 18f.

9. See Neale, *Technological Assessment*, 3–4 for a summary of the political roots of OTA. For a brief explanation of these questions see Westrum, *Technology and Society*, 335–36.

10. It is my opinion that this kind of irrational fear grips current Republican efforts to increase military spending to cope with real or imagined global or regional conflicts. With the fall of the Soviet Union, no credible military power exists in parity to the existing military might of America. This might is both mobile and can be fiercely focused, as we saw in Operation Desert Storm. These increases in military "pork" are proposed while many other budgets are being slashed *and* while former Speaker of the House Newt Gingrich promised to "put military spending on the table." The proposed increases signify not only an irrational fear but have led to more votes on the home front, especially for Representative James Hansen, R-Utah, and Senator James Imhoff, R-Oklahoma. See *U.S.A. Today*, "Porking up Defense," April 9, 1996, section C, p. 6.

11. A good introduction to the subject can be found in Marvin Waterstone, ed., *Risk And Society: The Interaction of Science, Technology and Public Policy* (Boston: Kluwer, 1992), 1–11.

12. For an excellent summary of this debate and a useful solution, see Kirstin Shrader-Frechette, "Public and Occupational Risk: The Double Standard, in Paul Durbin, ed., *Technology and Contemporary Life* (Boston: D. Reidel, 1988), 257–77. Compare Nicholas Rescher, *Risk: A Philosophical Introduction* (Washington, DC: University Press of America, 1983), 170–93.

13. See Lewis, *Technological Risk*, 30–33, for a fuller discussion of this topic.

14. This maxim does not hold for executives who knowingly suppress evidence that nicotine is addictive and, if ingested over long periods, therefore will cause health problems. Here the *knowing* maxim holds.

15. An excellent discussion of the effect of belief structures upon probability theories is W. D. Rowe, "The Spectrum of Uses of Risk Analysis" in Waterstone, ed., *Risk and Society*, 18–30. The best philosophical discussion on the relation of "belief" and science is Hendrick Hart, *Understanding Our World: An Integral Ontology* (New York: University Press of America, 1984), 325–370.

16. Lewis, *Technological Risk*, 69. Parenthesis in original.

17. An excellent example of this entire process is U.S. Government Printing Office, *Hearing Before the Subcommittee on Science, Space, and Technology* (Washington, DC: Government Printing Office, 1992), no. 53. For a history of the evolution of the assessment process in the context of the Federal Government, see Franklin P. Huddle, *Technology Information for Congress*, 3rd ed. (Washington, DC: Government Printing Office, 1979).

18. Lewis, *Technological Risk*, 81.

19. For an excellent exposition of the art of tradeoffs and its relationship to the notion of utility, see Edward Wenk, Jr., *Tradeoffs: Imperatives of Choice in a High-Tech World* (Baltimore: Johns Hopkins University Press, 1989), 44–58.

20. See the discussion in Lewis, *Technological Risk*, 147–53. Compare glossary terms in the General Accounting Office, *Health Risk Analysis*, 166–71, especially 166.

21. Lewis, *Technological Risk*, 148.

22. In a United States Court of Appeals case, the court ruled that a one in a nineteen billion exposure to risk constitutes a violation to this clause. Therefore, Orange Number 17 food dye had to be taken off the market.

23. A valuable examination of many of the issues that surround the Delaney Clause can be found in the General Accounting Office, *Health Risk Analysis*, 2–16.

24. Lewis, *Technological Risk*, 138–39.

25. Lewis, *Technological Risk*, 153.

26. Rudi Volti, *Society and Technological Change* (New York: St. Martin's, 1992), 237–46.

27. For a survey of the increasingly difficult task of determining public good, see W. D. Rowe, "Risk Analysis" in Waterstone, ed., *Risk and Society*, 24–28.

28. An comprehensive summary of common tradeoffs is found in Wenk, *Tradeoffs*, 205–6.

29. These two quotes appear in Kasper, *Technology Assessment*, 32. The phrase "happiest society we have ever known" points to technological optimism.

30. For a list of important subcommittees as well as a brief history of the Office of Science and Technology, see Kasper, *Technology Assessment*, 34, 35.

31. Richard Carpenter, "Technology Assessment and the Congress," in Kasper, *Technology Assessment*, 36. Emphasis in original.

32. Harold P. Green, "The Adversary Process in Technology Assessment," in Kasper, *Technology Assessment*, 49.

33. Note should be made of this point for subsequent analysis.

34. For a further development of this debate as it applies to risk assessment, see Steve Rayner, "Risk and Relativism in Science for Policy," in Johnson and Covello, *The Social and Cultural Construction of Risk*, (Boston: D. Reidel, 1987), 27–54. For an analysis and exposition of Positivism see Barbara MacKinnon, *American Philosophy: A Historical* (New York: State University, 1985), 508–65.

35. Rayner, "Risk and Relativism," 55.

36. For a further development of this debate as it applies to risk assessment see Rayner, "Risk and Relativism," 5–26.

37. A good summary of these specific reasons for minimizing risk is found in Johnson and Covello, *Social and Cultural Construction of Risk*, 27–54.

38. Johnson and Covello, *Social and Cultural Construction of Risk*, 28.

39. See Johnson and Covello, *Social and Cultural Construction of Risk*, 55–80.

40. A succinct summary of the "bursting boiler case" and other explicit examples can be found in Kasper, *Technology Assessment*, 75f.

41. See Kasper, *Technology Assessment*, 77. Compare A. R. Hall, *The Scientific Revolution, 1500–1800* (Boston: Beacon, 1956), 1–150, for the roots to this optimistic view of science.

42. For a detailed argument focusing on the connection between moral and technical optimism see my *Between God and Gold: Protestant Evangelicalism and the Industrial Revolution, 1820–1914* (Madison, NJ: Fairleigh Dickinson University Press, 1993), 27–115.

43. For a discussion of society and pluralism see Richard J. Mouw and Sander Griffioen, *Pluralisms and Horizons: An Essay in Christian Public Philosophy* (Michigan: William B. Eerdmans, 1993).

44. An excellent introduction to the relationship of the evaluation of technology to the formation of a world view is Carl Mitcham and Robert Mackey, eds., *Philosophy and Technology: Readings in the Philosophical Problems of Technology* (New York: Free Press, 1983). It is suggested that the reader pay close attention to the parts of the bibliography listed in the back of this book that have to do with the multidimensionality of technical assessment.

45. On the issue of the relationship between responsibility to technology and ethics, see Hans Jonas, *The Imperative of Responsibility: In Search of an Ethics for the Technological Age* (Chicago: University of Chicago Press, 1984), and Stephen V.

Monsma, *Responsible Technology: A Christian Perspective* (Grand Rapids, MI: William B. Eerdmans, 1986).

46. For the unacknowledged ethical commitments hidden in realism, see Frederick Ferre, *Philosophy of Technology* (Englewood, NJ: Prentice-Hall, 1988), 75–87.

47. See therefore Marvin Waterstone, ed., *Risk and Society*, 175–76.

48. For a discussion of how the ideal of progress effects our lives, see Bob Goudzwaard, *Capitalism and Progress: A Diagnosis of Western Society*, trans. and ed. Josina van Nuis Zylstra (Grand Rapids, MI: William B. Eerdmans, 1979). For a discussion on some of the foundations of modern technology, see Stephen H. Cutcliffe, et al., *New Worlds, New Technologies, New Issues: Research in Technology Studies, Volume 6* (Bethlehem, PA: Lehigh University Press, 1992), especially 168–85.

49. Waterstone, *Risk and Society*, 175–76.

50. Volti, *Society and Technological Change*, 238. Compare Duncan MacRae, Jr., "Science and the Formation of Policy in a Democracy," in Thomas J. Kuehn and Alan L. Porter, eds., *Science, Technology, and National Policy* (Ithaca: Cornell University Press, 1981), 497.

51. An excellent analysis of the "certitudinal" or faith aspect of science in general and risk analysis in particular is found in Kasper, *Technology Assessment*, 129–43.

52. See Peter Schouls, *Imposition of Method: A Study In Descartes and Locke* (Oxford: Clarendon University Press, 1980), for a careful analysis of how Descartes forced his mathematical-logical method upon the evidence and thus tainted it in his need to control the evidence.

53. Cutcliffe, et al., *New Worlds*, 42.

54. Cutcliffe, et al., *New Worlds*, 42–43. The reader may want to supplement this secondary source understanding with primary readings. On the Baconian ideal of liberation, see Francis Bacon, *The Great Instauration* and *New Atlantis*, ed. J. Weinberger (Arlington Heights, IL: Harlan Davidson, 1980); on Descartes, see René Descartes, *Discourse on Method*, trans. Lawrence J. Lafleur (Indianapolis: Bobbs-Merrill, 1956); on John Locke, *Treatise on Civil Government and a Letter Concerning Toleration*, ed. Charles Sherman (New York: Appleton, 1965).

55. Cutcliffe, et al., *New Worlds*, 44.

56. I would like to thank one of my students, Barry Bergling, for this insight.

57. An excellent treatise on the motif of control in Western technology is James Beniger, *The Control Revolution, Technological and Economic Origins of the Reformation Society* (Cambridge: Harvard University Press, 1986).

58. Cutcliffe, et al., *New Worlds*, 64.

59. See section in MacKinnon, *American Philosophy*, on tomes, p. 215–55, and on Dewey, p. 258–83, especially pages 224, 265, 266, 271, 279.

60. For a critique of this reductionism, see Ian G. Barbour, *Technology, Environment, and Human Values* (New York: Praeger, 1980), 170–75. See also Kenneth Boulding, *Economics as a Science* (New York: McGraw-Hill, 1970), 117–38.

61. Matthew 6:25c, *Oxford Study Edition: The New English Bible.*

62. For an expanded version of this argument, see Nicholas Rescher, *Risk: A Philosophical Introduction to the Theory of Risk Evaluation and Management* (Washington, DC: University Press of America, 1983), 179–81. Compare T. S. Schelling, "The Life You Save May Be Your Own," in *Problems in Public Expenditure Analysis*, ed. S. B. Caser, Jr. (Washington, DC: The Brookings Institution, 1966), 127–162.

63. For an introductory discussion of the "shadow market," see Ferre, *Technology*, 80–82 and Rescher, *Risk*, 135–40.

64. For a lingering view of the value-free nature of technology, see R. A. Buchanan, *Technology and Social Progress* (Oxford: Pergamon, 1985). Of note is the fact that the value-free argument was used by many who legitimated fission to the public.

65. Larry Rasmussen, "Mind Set and Moral Vision," in Frederick Ferre, ed., *Research in Philosophy and Technology* (Greenwich, CT: JAI Press, 1990), volume 10, 121.

66. A superb book that shows the relationship of religious belief to scientific investigation is Roy A. Clouser, *The Myth of Religious Neutrality: An Essay on the Hidden Role of Religious Belief in Theories* (Notre Dame, IN: Notre Dame University Press, 1991). Clouser first defines religion as that upon which all subsequent knowledge rests then shows how that even in mathematics, physics, and psychology, beliefs are foundationally at work.

67. See "The Postmodern Economy," in Stephen Cutcliffe, et al., *New Worlds*, 41–71; and Monsma, *Responsible Technology*, 77–102.

68. See Galileo, cited by Lewis Mumford in *Technics and Civilization* (New York: Harcourt, Brace and World, 1963), 40–52.

69. United States Environmental Protection Agency Science Advisory Board, *Reducing Risk: Setting Priorities and Strategies for Environmental Protection* (Washington, DC, Sept., 1990), 34.

70. See the testimony of Dr. S. Allen Heininger, President of the American Chemical Society's Center for Risk Management, in *Hearing Before the Subcommittee on the Environment*, One Hundred and Second Congress, First Session, May 21, 1991, (Washington DC: U.S. Government Printing Office, 1991), 191.

71. A critique of the arrogance of rationality can be found in Brynne, "Institutional Mythologies and Duel Societies in the Management of Risk," in Kunruether and Ley, *The Risk Analysis Controversy*, 127–133.

72. Elena Lugo, "New Dimensions for Action," in Stephen Cutcliffe, et al., *New Worlds*, 163. An able summary of what science may and may not attempt, and hence a critique of the arrogance of reason, is Peter Medawar's *The Limits of Science* (London: Oxford University Press, 1984).

73. Hart's treatment of Michael Polanyi, Gerard Radnitzky, and Thomas Kuhn in *Understanding Our World*, chapter 5 and the appendix, deserves careful reading and absorption. For a detailed and lucid discussion of how "commitment" and "belief" function in science, especially in the works of Descartes, Copernicus, and Tycho, see Marinus Dirk Stafleu, *Theories At Work: On the Structure and Functioning of Theories in Science* (New York: University Press of America, 1987), 216–60.

CHAPTER 4. TECHNOLOGICAL STRUCTURALIST

1. Egbert Schuurman, *Technology and the Future: A Philosophical Challenge*, trans. Herbert D. Morton (Toronto: Wedge, 1980), 327. Of course I am not saying that all the disciplines listed above have exactly the same view of law, or that Schumacher and Schuurman are similar on this point. I argue only that among competing

views of law, a common view is one that says that "natural law" delimits yet gives meaning to all of life. I mean by law a standard, principle, or ordinance that gives an action its character and limits.

2. Schuurman, *Technology and the Future*, 329–33. It must be noted that these aspects are made possible by the structuration within the creation by God, according to Schuurman. God gives, by means of the law, order, pattern, predictability. This order, in turn, leads to reality manifesting a certain level of meaning, purpose, and identity. Recognizing this fact is crucial for understanding Schuurman's structuralist position.

These aspects of reality were first philosophically delineated by the reformed school of thought known as Wijsbegeerte der Wetsidee, or the philosophy of the cosmonomic law idea. Schuurman's mentor, and one of the pioneers of this school, is Hendrik van Rissen, who was born in 1911 in the Netherlands. Van Rissen, like Schuurman, was both a practicing engineer and a philosopher. Van Rissen did his doctoral dissertation in 1949 on the interrelationship of philosophy and technology, many years before it became fashionable to think about this interrelationship.

3. Ernst F. Schumacher, *Small is Beautiful: Economics as if People Mattered* (New York: Harper and Row, 1973), 64. (Hereafter, *Beautiful*.)

4. Schumacher, *Small is Beautiful*, 146.

5. E. F. Schumacher, *A Guide for the Perplexed* (New York: Harper and Row, 1977), 19.

6. Schumacher, *Small is Beautiful*, 62.

7. Schumacher, *Small is Beautiful*, 159.

8. Schumacher, *Small is Beautiful*, 6.

9. The university that currently employs me sets the number at thirty students per class. Department budgets are cut if class size falls below this arbitrarily chosen figure. See *The Ball State Daily News*, volume 75, number 142 (1996), 1. One wonders if this same criterion is applied to administrators. That is, if the number of faculty are reduced does that mean that the standard of efficiency will used to reduce the number of administrators?

10. Schumacher, *Small is Beautiful*, 48.

11. Schumacher, *Small is Beautiful*, 139.

12. Schumacher, *A Guide for the Perplexed*, 12.

13. Schumacher, *A Guide for the Perplexed*, 19f. It remains to be seen if Schumacher's notion of "hierarchy" is only confined to complexity of levels of being, or if some kind of ontological chauvinism is present whereby the lower levels of being are considered inferior in quality to the higher levels.

14. Schumacher, *A Guide for the Perplexed*, 15–25.

15. Schumacher, *A Guide for the Perplexed*, 49. Emphasis in original.

16. Schumacher, *A Guide for the Perplexed*, 50–60.

17. Schumacher, *A Guide for the Perplexed*, 54–62.

18. Schumacher, *A Guide for the Perplexed*, 66.

19. Ergonomics is the study of the adaptation of technology to the biological, psychological, and social needs of the user.

20. Schumacher, *A Guide for the Perplexed*, 115–116.

21. Schumacher, *A Guide for the Perplexed*, 115–118.

22. The term *reductionism* means consequent shrinking of the complexity of reality to a manageable part, then the confusion of that limited part with the whole of reality. In almost all instances, this abstraction results in a truncated view of reality

in which the predetermined aspect of choice is believed to be central. Marx's economic materialism and optimism/pessimism aggrandizement of human freedom are perhaps the best examples of this reductionism and concomitant myopic worldview.

23. Schumacher, *A Guide for the Perplexed*, 134–35.

24. Schumacher, *A Guide for the Perplexed*, 135.

25. George McRobie, *Small is Possible*, foreword by Verna Schumacher (New York: Harper and Row, 1981). This book contains perhaps the best documentation of the many projects attempted under the banner of "small is beautiful." The foreword was written by Schumacher's widow.

26. See McRobie, *Small is Possible*, 1–13, for a detailed description of these points.

27. McRobie, *Small is Possible*, 185–86.

28. By "ecotechnical landscape" I mean to focus on machine and liquid capital as the primary tandem that empowers the production process. It is a consequence of the Industrial Revolution that machine and liquid capital have become so large and so intertwined or virtually synonymous with the production process.

29. McRobie, *Small is Possible*, 160.

30. McRobie, *Small is Possible*, 235–40.

31. McRobie, *Small is Possible*, 200.

32. McRobie, *Small is Possible*, 217.

33. McRobie, *Small is Possible*, 233–36.

34. E. F. Schumacher, *Good Work* (New York: Harper Collins, 1979), 27.

35. Schumacher, *Good Work*, 37.

36. Schumacher, *Good Work*, 79. It is common to hear faculty complain about poor pay. This complaint can be heard especially when during cuts, freezes, and insufficient raises, administrators raise their own salaries. One can only wonder how much more content and therefore productive we as faculty would be if more solidarity were present.

37. Schuurman was trained by a pioneer in the development of the philosophy of technology. From 1936–1943 van Riessen worked for a branch of the International Telephone and Telegraph Company developing cable systems. Van Riessen has occupied privately endowed chairs for Calvinistic philosophy at the Technical University of Delft (1951–1974) as well as the Technical University of Eindhoven (1961–1964). In 1964 he was appointed professor in systematic philosophy and philosophy of culture at the Free University of Amsterdam. He is now retired but still active.

A reasonably complete bibliography of van Riessen's work can be found in the article, "Symposium: Hendrik Van Riessen and Dutch Neo-Calvinist Philosophy of Technology," ed. and trans. Donald Morton, in *Research in Philosophy and Technology*, vol. 2 (London: JAI Press, 1979), 293–340. I am in debt to this article for my bibliographic comments on van Riessen.

38. Schuurman, *Technology and the Future*, 2.

39. On the relationship between ethics, system methodologies, and the scientific method see S. Strijbos, "Systems Methodologies for Managing our Technological Society: Toward a 'Disclosive Systems Thinking,'" unpublished paper.

40. Egbert Schuurman, *The Information Society: Impoverishment or Enrichment of Culture* (Potchefstroom, South Africa: Instituut vir Reformatoriese Studies, 1984), 12–13. Emphasis in original.

41. *Technology*, 5.

42. *Technology*, 40–1.

43. In Schuurman's earlier work, *Reflections on the Technological Society* (Toronto: Wedge, 1977), he argued that the fundamental ideological battle was between the "technocrats" (those espousing the use of more technology to cure all social ills—the optimists) and the "revolutionaries" (or those predominately neo-Marxists who threaten revolution because of the oppression of modern technology). It is currently difficult to speak of "revolutionaries" because such social radicalism dimmed after the 1970s. However much the pessimists resist modern technology in the name of freedom, technological enlargement has not been abated.

The strength of this work can be located in two areas. First, Schuurman begins his critique of both optimism and pessimism by showing their commonly shared basic commitments. Second, we see something of the history of both of these positions and how they come to become so deeply entrenched in culture.

44. I have argued, following Ellul's own designation, that his essential method is that of a "dialectic" or an absolutizing of two contradictory forces: freedom and technological necessity. Therefore, I would take exception to Schuurman's designation of Ellul as a transcendentalist. I believe that Ellul attempts to balance technological necessity with essential human freedom.

45. *Technology*, 54.

46. Egbert Schuurman, *Information Society*, 6.

47. For a deeper analysis of this theme, see Schuurman, *Reflections*, 25–63.

48. When Schuurman criticizes the religious principle of autonomy he is *not* criticizing the legitimate desire for freedom to chart one's destiny, at times apart from and yet with other trusted figures or institutions. Rather, he is criticizing the Enlightenment-inspired notion that people apart from God's laws or decrees can determine their own destinies and that of the earth by believing finally and chiefly in the healing effects of their own technological works. Compare *Technology*, 365–69 with *Reflections*, 10–14.

49. The term "word of God" is normally associated with the Bible. While Schuurman would not want to deny this part of God's word, he would argue that Christ and, here what has been called loosely general revelation and more loosely natural law refer to the fuller, more global sense of God's word.

50. The reformed Christian faith, historically rooted in the life and the work of John Calvin, spread to the Netherlands where it bore verdant cultural fruit. The crowning intellectual achievement was the founding of a major European university called the Free University that is located in Amsterdam.

51. Egbert Schuurman, "Crisis in Agriculture: A Philosophical Perspective on the Relation Between Agriculture and Nature," translated by Donald Morton, 9, in *Research in Philosophy and Technology, Volume 12* (London: Jai Press), 191–213.

52. One should note that Schuurman uses the phrase "fit in" in the last quote. For a similar *contextualist* view see Ian Barbour, *Ethics in an Age of Technology* (San Francisco: Harper, 1993). The strength of Barbour's work is his treatment of alternatives to current technological problems. His weakness is reflected in his inability to develop fully an *ontology* for his contextualism. He wants to call himself a contextualist, but he never develops a view of a context. This lack leaves his position superficial.

53. Schuurman, *Technology*, 331.

54. *Technology*, 329–30.

55. The relationship of sin to science is enormously complex. A brief explanation is now all that is possible. Schuurman is not saying that science equals sin.

Rather, he is saying that fragmentation is the result of autonomy. Scientific *abstraction* of one area or problem of life from its whole-life context for the purpose of study is the normative and necessary role that science plays in our essential task of understanding our world. However, science, like any other activity in life, can lead to fragmentation. Without understanding how its specialized knowledge is by definition related to the whole integral nature of reality, science can and does fall into sin or reductionism. See "Technology and Harvesting the Earth," 196–97. Compare *Technology*, 338–42.

56. Schuurman speaks about four levels of abstraction typically done in science. First there is the phase when discrete "facts" are split from creation's context and arranged into hypotheses. Second there is the abstraction of related facts and hypotheses from the first context and the making of a general principle. Third, and too often, there is the absolutization of one or more principles into a worldview component. An example of this would be the principle of relativity enlarged to a worldview known as relativism. Finally, there is the pretension that the scientist is a disinterested, value-free observer.

It is important to note these abstractions for two reasons. First, abstraction is a potential source of distortion according to Schuurman. Second, this tendency to distort must be kept in mind when we move to evaluating Schuurman. See "Technology and Harvesting the Earth," 196–97.

57. I affirm the multidimensionality of life but cannot agree with his particular characterizations of some of the principles of life. For example, his notion of efficiency as the norm for economics is tied to the neoclassical capitalistic paradigm, a view that he elsewhere rejects. In this paradigm one calls for *conservation* (which appears like the Christian notion of stewardship), even while one is maximizing inputs, consumption, and capital factors. Thus, an inherent tension between conservation and maximization is evident. This tension contradicts some of the basic assumptions of the philosophy Schuurman represents.

58. Egbert Schuurman, *Responsibility in the Technological Society* (n.d., n.p.). This draft-stage manuscript represents Schuurman's attempt to delineate more specifically what a responsible technology may look like. It was presented to me with the proviso that I do not take it as even approaching a completed manuscript. That said, I think this piece potentially represents his most insightful comments.

59. Egbert Schuurman, *Techniek: Middel of Moloch?* [Technology: Means or Moloch?] (Kampen: J. H. Kok, 1980). Moloch means "false god."

60. Egbert Schuurman, "Technology in a Christian-Philosophical Perspective," (Toronto: The Association for the Advancement of Christian Scholarship, 1979), 12–15.

61. Schuurman, "Technology in a Christian-Philosophical Perspective," 15.

62. *Responsibility*, 205f.

63. *Responsibility*, 210f.

64. Kelvin W. Willoughby, *Technological Choice: A Critique of the Appropriate Technology Movement* (San Francisco: Westview, 1990), 292. Willoughby, agreeing with Schumacher on this point, argues about

> the importance of attaining a good technological mix follows from the basic concept of technology fitting into reality and its corollaries. . . . The notion of a technological fit may not only apply to the individual technologies but to the whole *blend* of technologies employed within a community, 292. (Emphasis added.)

65. See especially Thomas Simon, "Appropriate Technology and Inappropriate Politics," in *Technology and Contemporary Life: Philosophy and Technology*, vol. 4, ed. Paul T. Durbin (Boston: D. Reidel, 1988), 108–9.

66. Economies of scale, in this instance, means that a given production may be controlled, or at least significantly influenced, by one large producer because that producer can supply the market at a lower per-unit cost. Accordingly, long-run average economic costs are lower while long-run social costs are higher because of the disruption caused by the large technical object. A case in point is nuclear production. Various forms of welfare and transfer payments are initiated, consequently, because employment is reduced to accommodate the natural monopoly.

67. Nicolas Jéquier, ed., *Appropriate Technology: Problems and Promises* (Paris: The Organization for Economic Cooperation and Development, 1976), 34–61.

68. Stanley P. Carpenter, "A Conversation Concerning Technology: The Appropriate Technology Movement," in *Appropriate Technology*, ed. Durbin, 87–88.

69. Simon, in *Appropriate Technology*, ed. Durbin, 109–11.

70. A member of Schuurman's own school of thought talks about the integrality of the world and hence a critique of mere specialized theoretical thought. See Hendrik Hart, *Understanding Our World: An Integral Ontology* (New York: University Press of America, 1984), 211–18.

71. Ian G. Barbour, *Ethics in an Age of Technology: The Gilfford Lectures, 1989–1991*, vol. 2. (San Francisco: Harper Publishers, 1993), 18.

72. See Hart, *Understanding Our World*, 346–45, and Daryl E. Chubin, et al., *Interdisciplinary Analysis and Research: Theory and Practice of Problem-Focused Research and Development* (Lomond, 1986), 9, 32, 96, 109, 157, 372–73, 462–63.

73. Instrumental rationality is that form of rationality or systematic thought that is characterized by expertise and specialization, and concentrates on one or a few disciplines or problems. It is "means" as opposed to "ends" dominated. It is analytic, specialized, and not holistic. I am not arguing that specialized thought is wrong; it is necessary. I am arguing that Schuurman manifests a form of rationality that complements, albeit inadequately, instrumental rationality. See in this regard, Joseph Weisenbaum, *Computer Power and Human Reason: From Judgement to Calculation* (New York: W. H. Freeman, 1976), 228–80; and Sander Griffioen and Jan Verhoogt, eds., for their "Introduction: Normativity and Contextuality in the Social Sciences," in *Norm and Context in the Social Sciences* (New York: University Press of America, 1990), 9–22.

74. Hendrik Hart and Kai Nielsen, *Searching for Community in a Withering Tradition: Conversations Between a Marxian Atheist and a Calvinian Christian* (New York: University Press of America, 1990), 190. Hart is a member of the same school of reformed philosophy as is Schuurman. Therefore, the quote and its implication seem all the more striking.

75. George Boas, *The Limits of Reason* (New York: Harper Brothers, 1961), 33–38. See also James Olthuis, "On World Views," in *Stained Glass: World Views and Social Science*, ed. Paul Marshall, Sander Griffioen, and Richard Mouw, (London: University of America Press, 1989), 26–40.

76. This is a German term used by the Amsterdam school of thought that refers to "object of thought." Thus, in theoretical gaze one is said to make a slice or a mode of reality an "object" of one's theoretical gaze. I have no quarrel with this analysis. I do not agree, however, with the often implicit Kantian assumption held at this point. Accordingly, when one makes a mode of life an object of one's gaze, one does not

NOTES TO CHAPTERS 4 AND 5

leave the realm of integral reality, a reality that must be accounted for *after* specialized thought has done its necessary work.

77. Thus Hart, *Understanding Our World*, 347, again, "The use of the term
meaning in this context highlights the dependence and relativity of all creaturely existence" [emphasis in original].

78. See, therefore, the philosopher of aesthetics Calvin Seerveld's *Rainbows for a
Fallen World* (Toronto: Tuppence Press, 1980).

79. A canvassing of Dutch sources will reveal the same ambivalent result: brilliant analysis but relatively superficial solutions. The following represent some additional Dutch works written by Schuurman followed by English title translations in
parenthesis:

————, *Filosofie van de Technische Wetenschappen* (*Philosophy of Technological Science*).
————, *Christenen in Babel* (*Christians in Babel*).
————, *De Kulturele Spanning Tussen* (Culture Between the Times).
————, *Na-denken over de Technisch-Vereniging* (Reflections on a Technologicl Society).
————, *Techniek en Toekomst* (Technology and the Future).
————, *Het 'Technische Paradijs'* (*The Technological Paradise.*)
————, *Tussen Technishe Overmacht en Menselijke Onmacht* (*Between Technological
Superiority and Human Powerlessness*).

I greatly appreciate the help of the library staff of the Free University in The
Netherlands who helped me gather this research during a recent sabbatical. Their
command of the English language greatly exceeded my understanding of Dutch.
Without this skill on their part, this research could not have been accomplished.

Chapter 5. Conclusion

1. I received at the conclusion of this manuscript a copy of Schuurman's latest
book, *Perspectives on Technology and Culture*, trans. John H. Kok (Sioux Center, IA:
Dordt College Press, 1995). This work does manifest *some* new alternatives. This is
especially true in the areas of energy and genetic manipulation. He suggests, for example, that a more normative energy policy would include the *complementary* use of
biomass, solar, water, and geothermal energy sources. Of course, conservation must
be increased. Nevertheless, the problem of alternatives presented with the rigor he
gives his critique is still missing.

2. Student review of this point was very skeptical. One student said that the de-
partmental system will never permit "lone rangers" enough room to make the kind of
connections mentioned. Another student said that no dean would ever allow as much
freedom as you imply because a bureaucracy will always follow money: "he who pays
the fiddler, calls the tune," was the student's protest. That day students retaught their
teacher a painful lesson I had learned but forgotten.

Glossary

The words are listed in this Glossary in the order of appearance within the text.

Ontological: The study of ontology focuses on different kinds of being and its attendant meaning and identity. I am using the term specifically to denote the structure for a many-sided reality, roughly paralleled by a liberal arts curriculum.

Autonomous: The word literally means "self-law". It refers to the belief that humans are their own final authority and that Reason specifically is the means to subdue nature to achieve desired ends.

Renaissance: The historical period 1300–1500 signaling the great medieval "rebirth" of classical art, literature, and especially learning. It centered on the belief that humans are the measure or the standard of all things.

Deism: A philosophical movement (1600–1700) that represents the belief in God on purely rational grounds. This God does not intervene in the daily operations of the universe. Rather, God is believed to have made the universe much like a watchmaker crafts a magnificent timepiece that subsequently runs autonomously. The entire universe, then, is said to run like a grand machine.

Heteronomous: The word literally means subject to an external authority or law that exists outside of one's Reason and consciousness. God's rule of the universe does intervene in our lives and therefore is relevant and authoritative. This view does not deny the use of reason or consciousness; it submits these faculties to heteronomous authority.

Industrial Revolution: was a dramatic intellectual, economic, and technological change in the nature of production. It started in England in 1760 and quickly spread to the United States and to Germany. It stressed large-scale, mass production and the centrality of the machine.

203

Capitalism: the owner of the means of production controls the economic system made-up of land, labor, and capital. This system claims to promote freedom of exchange. Its critics charge it with promoting a concentration of wealth and power. We capitalize the word in the text because it denotes a specific ideology, if not a religion.

Homo Faber: literally means, "man the tool-maker". Marx believed, as do many others, that the essential activity that characterized human identity was our tool making and using abilities.

Normative: A norm is a standard or a principle that is used in a foundational or a basic way. It serves to direct proper conduct. It does *not* mean normal.

Positivistic: The word is based on the philosophical movement known as Positivism. It was started by the nineteenth century thinker August Comte. This system of thought rejects speculation (especially religious) and founds its knowledge solely on observable, sense-based data and its relationships. Facts are said to originate from observation that is believed to be free from bias, evaluation, and dogmatism.

Reductionism: means the shrinking or the reducing of reality. Consequently one's view becomes narrow or myopic. Reducing the many aspects, sides, or "rooms" of creation to a concentration on one or a few is symptomatic of the problem of reductionism. Related to this term is . . .

Technicism: means the shrinking or the reducing of all of reality to the technical room or aspect. Technicism involves the absolutization of or the exaggeration of the meaning or the importance of technology to the point that the meaning and the reality of other non-technical areas of life are ignored, if not disappear.

Dialectic: means that two contradictory forces, words, or ideas create some tension that in turn brings about some result.

Technique: represents the social results of fabrication and manipulation. It is said to be autonomous, totalitarian, ecumenical, and determinative of our entire existence.

Neo-Marxism: The word "neo" means new. This newer kind of Marxist thought follows loosely in the intellectual footsteps of Karl Marx, the German nineteenth-century intellectual who is the father of Communism. Neo-Marxists stress that the current forces of oppression, excessive privilege, and power must be replaced by ones more democratic and freeing. New Marxists often resist violence as a means of freedom.

Desacralization: means to remove the sacred. Technique is thought to remove any thought of the holy, sacred, or supernatural in life.

Scylla and Charybdis: means being caught between two untenable perils. It is my understanding that Ellul forces us to choose two false and perilous alternatives: absolutized freedom and absolutized technique leave us with two dreadful choices.

Ideology: means the doctrines or foundational principles that form the basis of a worldview or an entire view of life.

Expertism: means the reign of experts or specially trained specialists whose knowledge and experience is thought to exceed any other's in a specific field of thought or endeavor.

Idolatry: is a theological term denoting the worship or whole-life service of a false god. In this case, technique is said falsely to claim ultimate allegiance.

Radical: means to the root. I am arguing that Ellul brings no radical or to the root critique to the problems presented by technique because he accepts some of the core assumptions of the position he is trying to critique.

Pluralistic forum: means a place in which many views are encouraged to be heard. I am suggesting that many different ideologies and religions be given official governmental room to express their opinions about the place and the limits of technology.

Absolutization: are the intellectual and the social processes whereby the importance of the use of one or a few aspects becomes paramount over the needs of other aspects of life. Technicism attempts to absolutize the technical room thereby causing a reduction in one's worldview. Economism, for example, exaggerates the place of the economic aspect for life.

Apocalyptic: refers to a revelation or a disclosure that is ultimately decisive for how we *should* live.

Bibliography

PRIMARY AND SECONDARY WORKS

Allen, Jonathan, ed. *March 4: Scientists, Students and Society*. Cambridge: MIT Press, 1970.

Bacon, Francis. *The Great Instauration*. Ed. J. Weinberger. Arlington Heights, IL: Harlan Davidson, 1980.

———. *New Atlantis*. Ed. J. Weinberger. Arlington Heights, IL: Harlan Davidson, 1980.

Barbour, Ian. *Ethics in An Age of Technology*. San Francisco: Harper and Row, 1994.

———. *Technology, Environment, and Human Values*. New York: Praeger, 1980.

Barnett, Richard S. and Ronald E. Muller. *Global Reach: The Power of Multinational Corporations*. New York: Simon and Schuster, 1974.

Beniger, James. *The Control Revolution, Technological and Economic Origins of the Reformation Society*. Cambridge: Harvard University Press, 1986.

Boas, George. *The Limits of Reason*. New York: Harper Brothers, 1961.

Bookchin, Michael. *The Ecology of Freedom: The Emergence and Dissolution of Hierarchy*. Palo Alto, CA: Cheshire, 1982.

———. *Toward an Ecological Society*. Montreal: Black Rose, 1980.

Boulding, Kenneth. *Economics as a Science*. New York: McGraw-Hill, 1970.

Buchanan, R. A. *Technology and Social Progress*. Oxford: Pergamon, 1985.

Burke, John G., and Marshall C. Eakin. *Theology and Change*. San Francisco: Boyd and Fraser, 1979.

Bury, J. B. *The Idea of Progress*. London: Macmillan, 1920.

Callahan, Raymond E. *Education and the Cult of Efficiency*. Chicago: University of Chicago Press, 1962.

Carr, Marilyn, ed. Introduction to *The A T Reader: Theory and Practice in Appropriate Technology*, by Frances Stewart. New York: Intermediate Technology Development Group of North America, 1985.

Christians, Clifford G., and Jay M. Van Hook, eds. *Jaques Ellul: Interpretive Essays*. Champaign: University of Illinois Press, 1981.

Chubin, Daryl E., Alan L. Porter, Frederick A. Rossini, and Terry Connolly, eds. *Interdisciplinary Analysis and Research: Theory and Practice of Problem-Focused Research and Development*. Mt. Airy, MD: Lomond, 1986.

Cutcliffe, Stephen H., Steven L. Goldman, Manuel Medina, and Jose Sanmartin, eds. *New Worlds, New Technologies, New Issues. Research in Technology Studies*, vol. 6. Bethlehem, PA: Lehigh University Press, 1992.

Descartes, Rene. *Discourse on Method*. Translated by Lawrence Lafleur. Indianapolis, IN: Bobbs-Merrill, 1956.

———. *Discourse on Method and the Meditations*. Trans. F. E. Sutcliffe. Harmondsworth, UK: Penguin, 1979.

Dooyeweerd, Herman. *Roots of Western Culture: Pagan, Secular, and Christian Options*. Toronto: Wedge, 1979.

Durbin, Paul T., ed. *Philosophy and Technology*. Vol. 7, *Broad and Narrow Interpretations of Philosophy of Technology*. Boston: Kluwer, 1990.

Ellul, Jacques. *The Ethics of Freedom*. Trans. and edited by Geoffrey W. Bromiley. Grand Rapids, MI: William B. Eerdmans, 1973.

———. "Mirror of These Ten Years." *Christian Century*, 18 February 1970. Wennemann, D. J., *Broad and Narrow Interpretations of the Philosophy of Technology*. Vol. 7. *An Interpretation of Jaques Ellul's Dialectical*. Ed. Paul T. Durbin. Boston: Kluwer Academic Publishers, 1990.

———. *The Technological Bluff*. Trans. Geoffrey W. Bromiley. Grand Rapids, MI: William B. Eerdmans, 1990.

———. *The Technological Society*. Translated by John Wilkinson. New York: Vintage, 1964.

———. *What I Believe*. Trans. Geoffrey W. Bromiley. Grand Rapids, MI: William B. Eerdmans, 1989.

Fairlie, Susan. "The Corn Laws and British Wheat Production." Quoted in Arnold Pacey, *The Culture of Technology*. Cambridge: MIT Press, 1984.

Ferre, Frederick. *Philosophy of Technology*. Englewood Cliffs, NJ: Prentice-Hall, 1988.

———. *Research in Philosophy of Technology*. Vol. 10. *Technology and Religion*. Greenwich, CT: JAI Press, 1990.

Fuller, Buckminster R. *No More Secondhand God and Other Writings*. Garden City, NY: Doubleday, 1963. Quoted in Frederick Ferre. *Philosophy of Technology*. Vol. 10. *Technology and Religion*. Greenwich, CT: JAI Press, 1990.

Goudzwaard, Bob. *Capitalism and Progress: A Diagnosis of Western Society*. Trans. and Ed. Tosina Van Nuis Zylstra. Grand Rapids, MI: William B. Eerdmans, 1979.

Habermas, Jürgen. *Technology and Science as "Ideology,"* 3rd ed. Frankfurt: Suhrkamp Verlag, 1969.

Hall, A. R. *The Scientific Revolution, 1500–1800*. Boston: Beacon, 1956.

Hart, Hendrick *Understanding Our World: An Integral Ontology*. New York: University Press of America, 1984.

———, and Kai Nielsen. *Searching for Community in a Withering Tradition: Conversations Between a Marxian Atheist and a Calvinian Christian*. New York: University Press of America, 1990.

Harvey, David. *The Condition of Postmodernity*. Oxford: Basil Blackwell, 1980.

Hopper, David H. *Technology, Theology, and the Idea of Progress*. Louisville, KY: John Knox Press, 1991.

Hoskins, W. G. "Harvest Fluctuations and English Economic History." Arnold Pacey. *The Culture of Technology*. Cambridge: MIT Press, 1984.

Huddle, Franklin P. *Technology Information for Congress*. 3rd edition. Washington, DC: Government Printing Office, 1979.

Jacob, Margaret C. *The Cultural Meaning of the Scientific Revolution*. New York: Alfred A. Knopf, 1988.

Jéquier, Nicolas, ed. *Appropriate Technology: Problems and Promises*. Paris: Organization for Economic Co-operation and Development, 1976.

Johnson, Branden B., and Vincent T. Covello. *The Social and Cultural Construction of Risk*. Boston: D. Reidel, 1987.

Jonas, Hans. *The Imperative of Responsibility: In Search of an Ethics for the Technological Age*. Chicago: University of Chicago Press, 1984.

Kahn, Herman, William Brown, Leon Martel, et al. *The Next Two Hundred Years: A Scenario for America and the World*. New York: Morrow, 1976. Quoted in Julian Simon. *The Ultimate Resource*. Princeton: Princeton University Press, 1981.

Kakar, Sudhir. *Frederick Taylor: A Study in Personality Innovation*. Cambridge: Harvard University Press, 1970.

Kasper, Raphael, G. *Technology Assessment: Understanding the Social Consequences of Technical Applications*. New York: Praeger, 1972.

Kimbrell, Andrew. *The Human Body Shop*. San Francisco: Harper Collins, 1994.

Klemm, Friedrich. *A History of Western Technology*. Translated by Dorothy Waley Singer. Cambridge: MIT Press, 1964.

Kunreuther, Howard C., and Eryl V. Ley, eds. *The Real Analysis Controversy: An Institutional Perspective*. New York: Springer-Verlag, 1987.

Landes, David S. *The Unbound Prometheus: Technological Change and Industrial Development in Western Europe from 1750 to the Present*. London: Cambridge University Press, 1969.

Lewis, H. W. *Technological Risk*. New York: Norton, 1990.

Locke, John. *Treatise on Civil Government and a Letter Concerning Toleration*. Ed. Charles Sherman. New York: Appleton, 1965.

Long, Franklin A. and Alexandra Olson, eds. *Appropriate Technology and Social Values—A Critical Appraisal*. Cambridge, MA: Ballinger, 1980.

Marcuse, Herbert. *An Essay On Liberation*. Boston: Beacon, 1969.

———. *One Dimensional Man*. Boston: Beacon, 1964.

Marx, Karl. *Das Kapital*. London: Eden and Cedar Paul, 1930.

MacLachlan, James. *Children of Prometheus: A History of Science and Technology*. Toronto: Wall and Emerson, 1989.

McRobie, George. *Small Is Possible*. Foreward by Verena Schumacher. New York: Harper and Row, 1981.

Medawar, Peter. *The Limits of Science*. London: Oxford University Press, 1984.

Mitcham, Carl, and Robert Mackey, eds. *Philosophy and Technology: Readings in the Philosophical Problems of Technology*. New York: Free Press, 1983.

Monsma, Stephen V. *Responsible Technology: A Christian Perspective*. Grand Rapids, MI: William B. Eerdmans, 1986.

Mumford, Lewis. *Technics and Civilization*. 1934; rpt., New York: Harcourt and Brace Company, 1963.

Neale, Ann. *Technological Assessment: Some Political and Theological Reflections*. Richmond, VA: Center for Theology and Public Policy, 1980.

Nielsen, Kai. *Research in Philosophy and Technology*. Vol. 1. *Technology as Ideology*. Greenwich, CT: JAI Press, 1978.

North, Douglas C. *Institutions, Institutional Change and Economic Performance*. New York: Cambridge University Press, 1990.

Pacey, Arnold. *The Culture of Technology*. Cambridge: MIT Press, 1984.

Polanyi, Michael. *Personal Knowledge*. Chicago: University of Chicago Press, 1958.

Rescher, Nicholas. *Risk: A Philosophical Introduction*. Washington, DC: University Press of America, 1983.

Rorty, Richard. *Consequences of Pragmatism*. Minneapolis: University of Minnesota Press, 1982.

Schouls, Peter. *Imposition of Method: A Study in Descartes and Locke*. Oxford: Clarendon University Press, 1980.

Schumacher, Ernest F. *Good Work*. New York: Harper Collins, 1979.

———. *A Guide for the Perplexed*. New York: Harper and Row, 1977.

———. *Small is Beautiful*. New York: Harper and Row, 1973.

Schuurman, Egbert. *Perspectives on Technology and Culture*. Trans. John H. Kok. Sioux Center, IA: Dordt College Press, 1995.

———. "Crisis in Agriculture: A Philosophical Perspective on the Relation Between Agriculture and Nature." In Donald Morton, ed. and trans. "Symposium: Hendrik can Riessen and Dutch Neo-Calvinist Philosophy of Technology." *Research in Philosophy and Technology*. Vol. 12. Greenwich, CT: JAI Press, 1979.

———. *The Information Society: Impoverishment or Enrichment of Culture*. Potchefstroom, South Africa: Instituut vir Reformatorese Studies, 1984.

———. *Reflections on the Technological Society*. Toronto: Wedge, 1977.

———. *Techniek: Middel or Moloch?* Kampen: J.H. Kok, 1980.

———. *Technology and the Future: A Philosophical Challenge*. Trans. Herbert Donald Morton. Toronto: Wedge, 1980.

———. *Technology in Christian-Philosophical Perspective*. Potchefstroom: Potchefstroom University for CHE, 1985.

Seerveld, Calvin. *Rainbows for a Fallen World*. Toronto: Tuppence Press, 1980.

Simon, Julian. *The Ultimate Resource*. Princeton: Princeton University Press, 1981.

Stafleu, Dirk. *Theories At Work: On the Structure and Functioning of Theories in Science*. New York: University Press of America, 1987.

Teich, Albert H. *Technology and the Future*, 5th ed. New York: St. Martin's, 1990.

Tillich, Paul. *The Spiritual Situation in Our Technological Society*. Edited and introduced by J. Mark Thomas. Georgia: Mercer University Press, 1988.

Turkle, Sherry. *The Second Self: Computers and the Human Spirit*. New York: Simon and Schuster, 1984.

Urdang, Laurence, ed. *The Random House Dictionary of the English Language: College Edition*. New York: Random House, 1968.

Vanderburg, William H., ed. *Perspectives on Our Age: Jaques Ellul Speaks on His Life and Works*. New York: Seabury Press, 1981. Quoted in D. J. Wennemann. *Broad and Narrow Interpretations of the Philosophy of Technology*. Vol. 7. *An Interpretation of Jaques Ellul's Dialectical*. Ed. Paul T. Durbin. Boston: Kluwer, 1990.

Van Riessen, H. *The Society of the Future*. Edited and translated by David H. Freeman. Philadelphia: Presbyterian and Reformed Publishing Company, 1953.

Volti, Rudi. *Society and Technological Change*. New York: St. Martin's, 1992.

Waterstone, Marvin, ed. *Risk and Society: The Interaction of Science, Technology and Public Policy*. Boston: Kluwer, 1992.

Wauzzinski, Robert. *Between God and Gold: Protestant Evangelicalism and the Industrial, 1820–1914*. Foreward by Martin Marty. Madison, NJ: Fairleigh Dickinson University Press, 1993.

Weisenbaum, Joseph. *Computer Power and Human Reason: From Judgement to Calculation*. New York: W. H. Freeman, 1976.

Wenk, Edward. *Tradeoffs: Imperatives of Choice in a High-Tech World*. Baltimore: Johns Hopkins University Press, 1989.

Wennemann, D. J. *Broad and Narrow Interpretations of the Philosophy of Technology*. Vol. 7. *An Interpretation of Jaques Ellul's Dialectical*. Ed. Paul T. Durbin. Boston: Kluwer, 1990.

Westrum, Ron. *Technology and Society: The Shaping of People and Things*. Wadsworth, 1991.

Wiener, Norbert. *Cybernetics, or Control and Communication in the Animal and the Machine*. Cambridge: MIT Press, 1950.

Willoughby, Kelvin W. *Technological Choice: A Critique of the Appropriate Technology Movement*. San Francisco: Westview, 1990.

Wilkinson, Loren, ed. *Earth Keeping: Christian Stewardship of Natural Resources*. Grand Rapids, MI: William B. Eerdmans,1983.

William, Charland A., Jr. *The Heart of the Global Village: Technology in the New Millennium*. Philadelphia: Trinity, 1990.

Winner, Langdon. *Autonomous Technology: Technics-Out-of-Control as a Theme in Political Thought*. Cambridge: MIT Press, 1977.

PUBLISHED ARTICLES AND MIMEOGRAPHS

Adams, Charles C. "Technology in a Christian Perspective." Toronto: Institute for Christian Studies, 1986.

Ayres, Clarence E. "The Industrial Way of Life." In *Technology and Change*, edited by John G. Burke and Marshall C. Eakin. San Francisco: Boyd and Fraiser, 1979.

Bellah, Robert. *Religion and the Technological Revolution in Japan and the United States*. Tempe, AZ: Department of Religious Studies, 1987.

Berdyaev, Nicholas. "Man and Machine." *The Bourgeois Mind and Other Essays*. New York: Sheed and Ward, 1934.

Birch, Charles. "Eight Fallacies of the Modern World and Five Axioms for a Postmodern World View." *Perspectives in Biology and Medicine* 1 (Autumn 1988): 12–30.

Byrd, Daniel, and Lester B. Lave. "A Framework for Risk Regulators." *Issues in Science and Technology* (Summer, 1987).

Carpenter, Richard. "Technology Assessment and the Congress." In *Technology Assessment: Understanding the Social Consequences of Technical Applications*, edited by Raphael G. Kasper. New York: Praeger, 1972.

Chelimsky, Eleanor. "Health Risk Analysis: Technical Adequacy in Three Selected Causes." Report to the Chairman Committee on Science, Space, and Technology, House of Representatives (September 1987).

Ezrahi, Yaron, Everett Mendelsohn, and Howard Segal. *Technology, Pessimism, and Postmodernism*. Hingham, MA: Kluwer, 1994.

Floreman, Samuel C. "In Praise of Technology." In *Technology and Change*, edited by John G. Burke and Marshall C. Eakin. San Francisco: Boyd and Fraser, 1979.

Green, Harold P. "The Adversary Process in Technology Assessment." In *Technology Assessment: Understanding the Social Consequences of Technical Applications*, edited by Raphael G. Kasper. New York: Praeger, 1972.

Griffioen, Sander, and Jan Verhoogt, eds. "Introduction: Normativity and Contextuality in the Social Sciences." In *Norm and Context in the Social Sciences*. New York: University Press of America, 1990.

Health Risk Analysis: Technical Adequacy Three Selected Cases. Washington, DC: General Accounting Office, 1987.

Hearing Before the Subcommittee on the Environment. Washington, DC: Government Printing Office, 1991.

Hearing Before the Subcommittee on Science, Space, and Technology. Washington, DC: Government Printing Office, 1992

Jaki, Stanley L. "The Three Faces of Technology." *The Intercollegiate Review* (Spring 1988): 37-46.

Legasov, Valeri, Leo Feoklistov, and Igor Kusmin. "Nuclear Power Engineering and International Security." *Soviet Life* 353, no. 2 (February 1986): 14.

Levin, Jack. "Unabomber A Hero to Some." *USA Today*, 11 April 1996, 1.

Lugo, Elena. "New Dimensions for Action." In *Research in Technology Studies*. Vol. 6, *New Worlds, New Technologies, New Issues*. Edited by Stephen H. Cutcliffe, Steven L. Goldman, Manuel Medina, and Jose Sanmartin. Bethlehem: Lehigh University Press, 1992.

MacRae, Duncan, Jr. "Science and the Formation of Policy in a Democracy." In *Science, Theology, and National Policy*, edited by Thomas J. Kuehn and Alan L. Porter. Ithaca: Cornell University Press, 1981.

Meier, Hugo A. "Technology and Democracy, 1800–1860." In *Technology and Change*, edited by John G. Burke and Marshall C. Eakin. San Francisco: Boyd and Fraser, 1979.

Morton, Donald, ed., trans. "Symposium: Hendrik van Riessen and Dutch Neo-Calvinist Philosophy of Technology." *Research in Philosophy and Technology*. Vol. 2. Greenwich, CT: JAI Press, 1979.

Neal, Ann. *Technological Assessment: Some Political and Theological Reflections*. Richmond, VA: Center for Theology and Public Policy, 1987.

Olthuis, James. "On World Views." In *Stained Glass: World Views and Social Science*, edited by Paul Marshall, Sander Griffioen, and Richard Mouw. London: University Press of America, 1989.

Radnitzky, Gerard. "Towards a Praxeological Theory of Research." *Systematics* 10 (1972): 131.

Rasmussen, Larry. "Mind Set and Moral Vision." In *Research in Philosophy of Technology*, vol. 10, edited by Frederick Ferre. *Technology and Religion*. Greenwich, CT: JAI Press, 1990.

Reducing Risk: Setting Priorities and Strategies for Environmental Protection. Washington, DC: United States Environmental Protection Agency Science Advisory Board, 1990.

Rowe, W. D. "The Spectrum of Uses of Risk Analysis." In *Risk and Society: The Interaction of Science, Technology and Public Policy*, edited by Marvin Waterstone. Boston: Kluwer, 1992.

Schelling, T. S. "The Life You Save May be Your Own." In *Problems in Public Expenditure Analysis*, edited by S. B. Case, Jr. Washington, DC: The Brookings Institution, 1966.

Schuurman, Egbert. "Technology in a Christian-Philosophical Perspective." Toronto: The Association for the Advancement of Christian Scholarship, 1979.

Shrader-Frechette, Kirstin. "Public and Occupational Risk: The Double Standard." In *Technology and Contemporary Life: Philosophy and Technology*, vol. 4, edited by Paul Durbin. Boston: D. Reidel, 1988.

Simon, Thomas. "Appropriate Technology and Inappropriate Politics." In *Technology and Contemporary Life: Philosophy and Technology*, vol. 4, edited by Paul T. Durbin. Boston: D. Reidel, 1988.

Unpublished Works

Ablelson, Philip H. "Enough of Pessimism: Insights into the Imperatives of Science and the Modern World." Unpublished paper.

Frost, Taggart. "Attitudinal Dispositions Toward Technology and Religiosity." Brigham Young University, 1981.

Prall, James W. "Computer and Natural Language: Will They Find Happiness Together? Master of Philosophical Foundations, Institute of Christian Studies, 1985.

Schuurman, Egbert. "Responsibility in the Technological Society." Unpublished paper.

———. "The Technological Culture Between the Times: A Christian Philosophical Assessment of Contemporary." Unpublished paper.

Strijbos, S. "Systems Methodologies for Managing Our Technological Society: Towards a Disclosive Systems Thinking." Unpublished paper.

Index

213